Renewable Energy Made Simple

A Beginner's Guide to Clean Power

Maggie Salazar

Renewable Energy Made Simple

*© Copyright 2024 by **Maggie Salazar***

All rights reserved

This document is geared towards providing exact and reliable information with regards to the topic and issue covered. The publication is sold with the idea that the publisher is not required to render accounting, officially permitted, or otherwise, qualified services. If advice is necessary, legal or professional, a practiced individual in the profession should be ordered.

From a Declaration of Principles which was accepted and approved equally by a Committee of the American Bar Association and a Committee of Publishers and Associations.

In no way is it legal to reproduce, duplicate, or transmit any part of this document in either electronic means or in printed format. Recording of this publication is strictly prohibited and any storage of this document is not allowed unless with written permission from the publisher. All rights reserved.

The information provided herein is stated to be truthful and consistent, in that any liability, in terms of inattention or otherwise, by any usage or abuse of any policies, processes, or directions contained within is the solitary and utter responsibility of the recipient reader. Under no circumstances will any legal responsibility or blame be held against the publisher for any reparation, damages, or monetary loss due to the information herein, either directly or indirectly.

Respective authors own all copyrights not held by the publisher.

The information herein is offered for informational purposes solely, and is universal as so. The presentation of the information is without contract or any type of guarantee assurance.

The trademarks that are used are without any consent, and the publication of the trademark is without permission or backing by the trademark owner. All trademarks and brands within this book are for clarifying purposes only and are owned by the owners themselves, not affiliated with this document.

TABLE OF CONTENTS

Chapter 1: Types of Renewable Energy Sources

Solar Energy: Harnessing the Power of the Sun

Solar energy, a cornerstone of renewable energy sources, offers a promising solution to the world's growing energy demands. The sun, a colossal nuclear reactor, emits energy in the form of light and heat, which can be harnessed to generate electricity and provide heating. This chapter delves into the intricacies of solar energy, exploring its potential, technology, and applications.

The journey of solar energy begins with the sun's rays reaching the Earth. Photovoltaic (PV) cells, the heart of solar panels, capture this sunlight and convert it into electricity. These cells are made from semiconductor materials, typically silicon, which absorb photons and release electrons, creating an electric current. The efficiency of PV cells has improved significantly over the years, making solar power a viable option for both residential and commercial use.

One of the most appealing aspects of solar energy is its abundance. The sun provides more energy in an hour than the entire world consumes in a year. This vast potential makes solar energy an attractive option for reducing reliance on fossil fuels and decreasing greenhouse gas emissions. Moreover, solar power systems can be installed on rooftops, in open fields, or even integrated into building materials, offering flexibility in deployment.

The adoption of solar energy is not without its challenges. The initial cost of solar panels and installation can be a barrier for many homeowners. However, the long-term savings on electricity bills and the availability of government incentives and tax credits can offset these costs. Additionally, advancements in technology have led to a decrease in the price of solar panels, making them more accessible to a wider audience.

Another consideration is the intermittency of solar energy. The sun does not shine at night, and weather conditions can affect the amount of sunlight reaching the panels. To address this issue, energy storage solutions, such as batteries, are essential. These systems store excess energy generated during sunny periods for use during cloudy days or at night, ensuring a consistent power supply.

Solar energy is not limited to electricity generation. Solar thermal systems harness the sun's heat for various applications, including water heating, space heating, and even cooling. These systems use collectors to absorb solar radiation and transfer the heat to a fluid, which is then used to heat water or air. Solar thermal technology is particularly effective in regions with high solar insolation, where it can significantly reduce energy consumption for heating purposes.

The environmental benefits of solar energy are substantial. Unlike fossil fuels, solar power does not produce air pollutants or carbon dioxide during operation. This clean energy source contributes to improved air quality and helps combat climate change. Furthermore, solar panels have a long lifespan, often exceeding 25 years, and require minimal maintenance, making them a sustainable choice for energy production.

In addition to environmental advantages, solar energy can enhance energy independence. By generating electricity locally, communities can reduce their reliance on imported fuels and increase their resilience to energy price fluctuations. This decentralization of energy production also reduces transmission losses, as electricity is consumed closer to where it is generated.

The integration of solar energy into the grid presents both opportunities and challenges. As the share of solar power increases, grid operators must manage the variability and ensure a stable supply of electricity. Advanced grid technologies, such as smart grids and demand response systems, can help balance supply and demand, facilitating the integration of solar energy.

Innovations in solar technology continue to drive the industry forward. Bifacial solar panels, which capture sunlight on both sides, offer increased efficiency and energy yield. Building-integrated photovoltaics (BIPV) incorporate solar cells into building materials, such as windows and facades, seamlessly blending energy generation with architecture. These advancements, along with ongoing research into new materials and manufacturing processes, promise to further enhance the performance and affordability of solar energy systems.

The global adoption of solar energy is on the rise, with countries around the world investing in large-scale solar projects. These initiatives not only contribute to energy security but also create jobs and stimulate economic growth. As the cost of solar technology continues to decline, it is expected that solar energy will play an increasingly significant role in the global energy mix.

For individuals considering solar energy, understanding the basics of solar panel technology and assessing their home's solar potential are crucial steps. Factors such as roof orientation, shading, and local climate can impact the efficiency of a solar system. Consulting with a professional installer can provide valuable insights and help determine the most suitable system for a specific location.

Financial considerations are also important when transitioning to solar energy. While the upfront costs can be substantial, various financing options, such as solar loans and power purchase agreements, can make the investment more manageable. Additionally, government incentives, rebates, and tax credits can significantly reduce the overall cost, making solar energy an attractive option for homeowners and businesses alike.

The installation and maintenance of solar systems require careful planning and execution. Proper installation ensures optimal performance and longevity, while regular maintenance, such as cleaning and inspection, can prevent potential issues and maintain efficiency. Working with experienced professionals can ensure that the system is installed correctly and operates at peak performance.

Solar energy represents a transformative force in the quest for sustainable energy solutions. Its potential to provide clean, abundant, and affordable power makes it a key player in the transition to a low-carbon future. By harnessing the power of the sun, we can reduce our environmental impact, enhance energy security, and pave the way for a more sustainable world.

Wind Energy: Capturing the Breeze

Wind energy, a vital component of the renewable energy landscape, harnesses the kinetic power of the wind to generate electricity. This chapter delves into the mechanics, benefits, and challenges of wind energy, offering insights into how this ancient force of nature is being utilized in modern times to meet the growing demand for clean energy.

The concept of capturing wind energy is not new. For centuries, humans have used windmills for tasks such as grinding grain and pumping water. Today, the technology has evolved into sophisticated wind turbines that convert wind's kinetic energy into electrical power. These turbines, often towering over landscapes, consist of blades, a rotor, a nacelle, and a tower. As the wind blows, it turns the blades, which spin the rotor connected to a generator inside the nacelle, producing electricity.

One of the most compelling aspects of wind energy is its sustainability. Wind is an inexhaustible resource, constantly replenished by natural atmospheric processes. Unlike fossil fuels, wind energy does not produce greenhouse gases or air pollutants during operation, making it an environmentally friendly option. Additionally, wind farms can be established onshore or offshore, providing flexibility in site selection and allowing for the optimization of wind resources.

The efficiency of wind energy depends on several factors, including wind speed, turbine design, and site location. Wind speed is crucial, as the power generated by a turbine increases exponentially with wind velocity. Therefore, selecting sites with consistent and strong winds is essential for maximizing energy output. Coastal areas, open plains, and hilltops are often ideal locations for wind farms due to their favorable wind conditions.

Despite its advantages, wind energy faces several challenges. One of the primary concerns is the intermittency of wind. Wind patterns can be unpredictable, leading to fluctuations in energy production. To address this issue, wind energy is often integrated with other renewable sources or supported by energy storage systems to ensure a stable power supply. Additionally, advancements in forecasting technology have improved the ability to predict wind patterns, aiding in grid management and planning.

The visual and environmental impact of wind turbines is another consideration. While wind farms occupy relatively small land areas, their presence can alter landscapes and affect local wildlife. Birds and bats, in particular, can be at risk of collision with turbine blades. To mitigate these impacts, careful site selection, turbine design improvements, and wildlife monitoring programs are implemented. Moreover, community engagement and transparent communication are vital in addressing public concerns and fostering acceptance of wind projects.

Wind energy offers significant economic benefits. The construction and maintenance of wind farms create jobs and stimulate local economies. Furthermore, wind energy can contribute to energy independence by reducing reliance on imported fuels and stabilizing energy prices. As technology advances and economies of scale are realized, the cost of wind energy continues to decrease, making it increasingly competitive with traditional energy sources.

For individuals interested in harnessing wind energy on a smaller scale, residential wind systems present an opportunity. These systems, typically consisting of a small turbine mounted on a tower, can provide supplemental power for homes or small businesses. Before investing in a residential wind system, it is essential to evaluate the wind potential in the area, considering factors such as average wind speed, zoning regulations, and potential obstructions like trees or buildings.

The installation of a wind turbine requires careful planning and consideration. Factors such as tower height, turbine size, and site location must be assessed to ensure optimal performance. Additionally, understanding local regulations and obtaining necessary permits are crucial steps in the process. Working with experienced professionals can help navigate these complexities and ensure a successful installation.

Maintenance is an important aspect of wind energy systems. Regular inspections and servicing are necessary to keep turbines operating efficiently and to prevent potential issues. While wind turbines are generally reliable, components such as blades, bearings, and electrical systems may require attention over time. Establishing a maintenance schedule and working with qualified technicians can help extend the lifespan of the system and maximize energy production.

The future of wind energy is promising, with ongoing research and innovation driving the industry forward. Technological advancements, such as larger and more efficient turbines, floating offshore wind farms, and improved grid integration, are expanding the potential of wind energy. As the world seeks to transition to a low-carbon economy, wind energy is poised to play a crucial role in meeting global energy needs sustainably.

Wind energy represents a powerful tool in the fight against climate change, offering a clean, renewable, and increasingly cost-effective source of electricity. By capturing the breeze, we can reduce our environmental impact, enhance energy security, and contribute to a more sustainable future. As technology continues to advance and the benefits of wind energy become more widely recognized, its role in the global energy landscape is set to grow, providing a vital pathway to a cleaner, greener world.

Hydropower: Energy from Water

Hydropower, one of the oldest and most reliable sources of renewable energy, harnesses the power of moving water to generate electricity. This chapter delves into the mechanics, benefits, and challenges of hydropower, offering insights into how this natural force is being utilized to meet the growing demand for sustainable energy.

The principle behind hydropower is simple yet effective. Water, driven by gravity, flows from higher elevations to lower ones, and this movement can be captured to produce energy. The most common form of hydropower involves the construction of dams on large rivers. These dams create reservoirs that store vast amounts of water. When released, the water flows through turbines, spinning them to generate electricity. The process is efficient and can produce large quantities of power, making it a cornerstone of many national energy strategies.

One of the most significant advantages of hydropower is its ability to provide a consistent and reliable source of electricity. Unlike solar or wind energy, which depend on weather conditions, hydropower can generate electricity continuously, as long as there is a sufficient water supply. This reliability makes it an excellent option for meeting base-load energy demands and providing grid stability.

Hydropower also offers substantial environmental benefits. It produces no direct emissions of greenhouse gases or air pollutants, contributing to cleaner air and a reduction in climate change impacts. Additionally, hydropower facilities can be designed to support local ecosystems by creating habitats for fish and wildlife. However, the construction of large dams can have significant environmental and social impacts, including habitat disruption, changes in water quality, and displacement of communities. To mitigate these effects, modern hydropower projects often incorporate environmental assessments and community engagement processes to ensure sustainable development.

The economic benefits of hydropower are considerable. Once a hydropower plant is constructed, the operational and maintenance costs are relatively low, and the facilities can have long lifespans, often exceeding 50 years. This longevity, combined with the ability to generate large amounts of electricity, makes hydropower a cost-effective energy source. Furthermore, hydropower projects can stimulate local economies by creating jobs and supporting infrastructure development.

In addition to large-scale hydropower, small-scale or micro-hydropower systems offer opportunities for decentralized energy production. These systems are particularly suitable for rural or remote areas where access to the central grid is limited. Micro-hydropower systems can be installed on small streams or rivers, providing a sustainable and independent energy source for communities. The installation of such systems requires careful planning and consideration of local environmental conditions, but they can offer significant benefits in terms of energy security and community empowerment.

The integration of hydropower into the energy grid presents both opportunities and challenges. Hydropower plants can provide flexible and responsive power generation, making them ideal for balancing supply and demand fluctuations. This capability is particularly valuable as the share of intermittent renewable energy sources, such as wind and solar, increases. However, the management of water resources is crucial to ensure that hydropower plants can operate effectively without compromising other water uses, such as agriculture, drinking water supply, and ecosystem health.

Innovations in hydropower technology continue to enhance its potential. Advances in turbine design, such as fish-friendly turbines, aim to reduce the environmental impact of hydropower facilities. Additionally, the development of pumped-storage hydropower offers a solution for energy storage, allowing excess electricity to be stored and released when needed. This technology can help integrate renewable energy sources into the grid by providing a reliable backup during periods of low generation.

For individuals interested in exploring hydropower, understanding the basics of hydroelectric power generation and assessing the potential of local water resources are essential steps. Factors such as water flow, head height, and environmental considerations must be evaluated to determine the feasibility of a hydropower project. Consulting with experts and conducting thorough site assessments can provide valuable insights and help identify the most suitable solutions for specific locations.

The future of hydropower is promising, with ongoing research and innovation driving the industry forward. As the world seeks to transition to a low-carbon economy, hydropower is poised to play a crucial role in meeting global energy needs sustainably. By harnessing the power of water, we can reduce our environmental impact, enhance energy security, and contribute to a more sustainable future. As technology continues to advance and the benefits of hydropower become more widely recognized, its role in the global energy landscape is set to grow, providing a vital pathway to a cleaner, greener world.

Biomass: Organic Material as Fuel

Biomass energy, derived from organic materials, stands as a versatile and sustainable source of power. This chapter delves into the intricacies of biomass, exploring its potential to transform waste into valuable energy while addressing environmental concerns and offering practical insights for beginners.

At its core, biomass energy is generated from organic materials such as plant matter, agricultural residues, and even animal waste. These materials, rich in stored solar energy, can be converted into heat, electricity, or biofuels through various processes. The versatility of biomass lies in its ability to utilize a wide range of feedstocks, making it an adaptable solution for diverse energy needs.

One of the most common methods of harnessing biomass energy is through combustion. In this process, organic materials are burned to produce heat, which can then be used directly for heating or to generate electricity through steam turbines. This method is particularly effective for large-scale power generation, where biomass can be co-fired with coal in existing power plants, reducing carbon emissions and reliance on fossil fuels.

Another promising avenue for biomass energy is anaerobic digestion. This biological process involves the breakdown of organic matter by microorganisms in the absence of oxygen, producing biogas—a mixture of methane and carbon dioxide. Biogas can be used as a renewable substitute for natural gas, providing a clean and efficient energy source for heating, electricity generation, and even transportation. The digestate, a byproduct of this process, can be used as a nutrient-rich fertilizer, closing the loop in a sustainable agricultural cycle.

Biomass can also be converted into liquid biofuels, such as ethanol and biodiesel, through biochemical processes. Ethanol, typically produced from sugarcane or corn, is used as a renewable fuel additive in gasoline, reducing greenhouse gas emissions and enhancing energy security. Biodiesel, derived from vegetable oils or animal fats, can be used in diesel engines with little or no modification, offering a sustainable alternative to petroleum-based diesel.

The environmental benefits of biomass energy are significant. By utilizing organic waste materials, biomass energy reduces the amount of waste sent to landfills, mitigating methane emissions and conserving valuable landfill space. Additionally, biomass energy is considered carbon-neutral, as the carbon dioxide released during combustion is offset by the carbon dioxide absorbed by plants during their growth. This balance makes biomass a key player in efforts to combat climate change and transition to a low-carbon economy.

Despite its advantages, biomass energy faces several challenges. The collection, transportation, and storage of biomass feedstocks can be logistically complex and costly. Ensuring a consistent and reliable supply of biomass materials is crucial for the viability of biomass energy projects. Furthermore, the sustainability of biomass production must be carefully managed to avoid negative impacts on land use, biodiversity, and food security.

For individuals interested in exploring biomass energy, understanding the types of feedstocks available and their potential applications is essential. Agricultural residues, such as straw and corn stover, offer a readily available source of biomass, while dedicated energy crops, like switchgrass and miscanthus, can be cultivated specifically for energy production. Additionally, organic waste from food processing, forestry, and municipal sources can be utilized, turning waste into a valuable resource.

The implementation of biomass energy systems requires careful planning and consideration. Factors such as feedstock availability, conversion technology, and end-use applications must be evaluated to determine the most suitable approach. Consulting with experts and conducting feasibility studies can provide valuable insights and help identify the best solutions for specific locations and needs.

The integration of biomass energy into existing energy systems presents both opportunities and challenges. Biomass can complement other renewable energy sources, providing a stable and dispatchable power supply. However, the development of biomass energy projects must be balanced with environmental and social considerations to ensure sustainable outcomes.

Innovations in biomass technology continue to enhance its potential. Advances in gasification, pyrolysis, and torrefaction offer new pathways for converting biomass into energy, improving efficiency and expanding the range of feedstocks that can be utilized. Additionally, research into algae-based biofuels holds promise for producing high-energy-density fuels with minimal land and water requirements.

The future of biomass energy is promising, with ongoing research and innovation driving the industry forward. As the world seeks to transition to a low-carbon economy, biomass energy is poised to play a crucial role in meeting global energy needs sustainably. By harnessing the power of organic materials, we can reduce our environmental impact, enhance energy security, and contribute to a more sustainable future. As technology continues to advance and the benefits of biomass energy become more widely recognized, its role in the global energy landscape is set to grow, providing a vital pathway to a cleaner, greener world.

Geothermal Energy: Heat from the Earth

Geothermal energy, a powerful and consistent source of renewable energy, taps into the Earth's internal heat to generate electricity and provide heating solutions. This chapter explores the potential of geothermal energy, its applications, and the practical steps for harnessing this subterranean resource.

The Earth's core, a molten mass of rock and metal, generates immense heat. This heat radiates outward, warming the surrounding layers of rock and water. Geothermal energy exploits this natural heat, which can be accessed through various geological formations such as hot springs, geysers, and volcanic regions. The most common method of harnessing geothermal energy involves drilling wells into geothermal reservoirs to extract steam or hot water, which is then used to drive turbines and generate electricity.

One of the most compelling advantages of geothermal energy is its reliability. Unlike solar or wind energy, which are subject to weather fluctuations, geothermal energy provides a constant and stable power supply. This consistency makes it an excellent option for base-load electricity generation, ensuring a steady flow of power to the grid. Additionally, geothermal power plants have a small land footprint compared to other renewable energy sources, minimizing their impact on the surrounding environment.

Geothermal energy is not limited to electricity generation. It can also be used for direct heating applications, such as district heating systems, greenhouse heating, and industrial processes. In residential settings, geothermal heat pumps offer an efficient way to heat and cool homes by transferring heat between the ground and the building. These systems take advantage of the relatively constant temperature of the Earth's surface, providing a sustainable and cost-effective solution for climate control.

The environmental benefits of geothermal energy are significant. Geothermal power plants produce minimal greenhouse gas emissions compared to fossil fuel-based power plants, contributing to cleaner air and a reduction in climate change impacts. Furthermore, geothermal energy is considered a renewable resource, as the Earth's heat is continuously replenished by natural processes. This sustainability makes geothermal energy a key player in efforts to transition to a low-carbon economy.

Despite its advantages, geothermal energy faces several challenges. The availability of geothermal resources is geographically limited, with the most accessible and economically viable sites located near tectonic plate boundaries or volcanic regions. This limitation can restrict the widespread adoption of geothermal energy in certain areas. Additionally, the initial cost of drilling and developing geothermal wells can be high, although the long-term operational costs are relatively low.

For individuals interested in exploring geothermal energy, understanding the local geological conditions and resource potential is essential. Conducting thorough site assessments and feasibility studies can provide valuable insights into the viability of geothermal projects. Consulting with experts and leveraging existing geological data can help identify the most suitable locations for geothermal development.

The integration of geothermal energy into existing energy systems presents both opportunities and challenges. Geothermal power plants can provide flexible and responsive power generation, making them ideal for balancing supply and demand fluctuations. However, the development of geothermal projects must be balanced with environmental and social considerations to ensure sustainable outcomes.

Innovations in geothermal technology continue to enhance its potential. Advances in drilling techniques, such as enhanced geothermal systems (EGS), aim to expand the range of accessible geothermal resources by creating artificial reservoirs in hot dry rock formations. Additionally, research into low-temperature geothermal applications holds promise for utilizing geothermal energy in regions with lower heat gradients.

The future of geothermal energy is promising, with ongoing research and innovation driving the industry forward. As the world seeks to transition to a low-carbon economy, geothermal energy is poised to play a crucial role in meeting global energy needs sustainably. By harnessing the Earth's heat, we can reduce our environmental impact, enhance energy security, and contribute to a more sustainable future. As technology continues to advance and the benefits of geothermal energy become more widely recognized, its role in the global energy landscape is set to grow, providing a vital pathway to a cleaner, greener world.

Chapter 2: Getting Started with Solar Power

Basics of Solar Panel Technology

Solar panel technology, a cornerstone of modern renewable energy solutions, has revolutionized the way we harness the sun's power. Understanding the basics of this technology is essential for anyone interested in adopting solar energy, whether for residential, commercial, or industrial purposes. This chapter delves into the fundamental components, workings, and considerations of solar panel systems, providing a comprehensive guide for beginners.

At the heart of solar panel technology are photovoltaic (PV) cells, which are responsible for converting sunlight into electricity. These cells are typically made from silicon, a semiconductor material that exhibits the photovoltaic effect. When sunlight strikes the surface of a PV cell, it excites electrons, creating an electric current. This direct current (DC) is then converted into alternating current (AC) by an inverter, making it suitable for use in homes and businesses.

Solar panels are composed of multiple PV cells connected in series and encapsulated within protective layers of glass and plastic. These panels are mounted on rooftops, ground-mounted structures, or integrated into building materials, depending on the specific application and site conditions. The efficiency of a solar panel, which determines how much sunlight is converted into usable electricity, is influenced by factors such as cell type, panel design, and environmental conditions.

There are several types of solar panels available, each with its own advantages and limitations. Monocrystalline panels, known for their high efficiency and sleek appearance, are made from single-crystal silicon and are ideal for installations with limited space. Polycrystalline panels, made from multiple silicon crystals, offer a more cost-effective solution with slightly lower efficiency. Thin-film panels, which use a variety of materials such as cadmium telluride or amorphous silicon, are lightweight and flexible, making them suitable for unconventional applications.

The performance of solar panels is affected by several environmental factors. Sunlight intensity, or irradiance, plays a crucial role in determining the amount of electricity generated. Panels perform best under direct sunlight, but they can still produce power on cloudy days, albeit at reduced efficiency. Temperature also impacts performance, as high temperatures can decrease the efficiency of PV cells. Proper ventilation and installation techniques can help mitigate this effect.

Shading is another critical consideration when installing solar panels. Even partial shading from trees, buildings, or other obstructions can significantly reduce a panel's output. To address this issue, microinverters or power optimizers can be used to maximize the performance of individual panels within a larger array. These devices ensure that shading on one panel does not disproportionately affect the entire system's output.

The orientation and tilt of solar panels are essential for optimizing energy production. In the northern hemisphere, panels should generally face south to capture the maximum amount of sunlight throughout the day. The tilt angle should be adjusted based on the latitude of the installation site to achieve optimal performance. Some systems incorporate tracking mechanisms that adjust the panel's position to follow the sun's path, further enhancing energy capture.

The installation of solar panels involves several key steps, beginning with a thorough site assessment. This assessment evaluates factors such as roof condition, available space, shading, and local regulations. Once the site is deemed suitable, the design and engineering phase involves selecting the appropriate panel type, inverter, and mounting system. Professional installation ensures that the system is securely mounted, properly wired, and compliant with safety standards.

Financial considerations play a significant role in the decision to adopt solar energy. While the initial cost of solar panels and installation can be substantial, various financing options are available to make the investment more accessible. Solar loans, leases, and power purchase agreements (PPAs) offer different pathways to ownership, allowing individuals to choose the option that best suits their financial situation. Additionally, government incentives, rebates, and tax credits can significantly reduce the overall cost, making solar energy an attractive option for many.

The maintenance of solar panels is relatively straightforward, as they have no moving parts and are designed to withstand harsh weather conditions. Regular cleaning to remove dust, dirt, and debris can help maintain optimal performance. Periodic inspections by qualified professionals can identify potential issues, such as loose connections or damaged components, ensuring the system operates efficiently over its lifespan.

The benefits of solar panel technology extend beyond environmental considerations. By generating electricity on-site, individuals and businesses can reduce their reliance on the grid, leading to lower energy bills and increased energy independence. Solar energy also contributes to a reduction in greenhouse gas emissions, supporting efforts to combat climate change and promote sustainability.

As technology continues to advance, the efficiency and affordability of solar panels are expected to improve, further driving their adoption. Innovations in materials, manufacturing processes, and system integration are expanding the possibilities for solar energy, making it a key player in the transition to a clean energy future.

Understanding the basics of solar panel technology empowers individuals to make informed decisions about adopting solar energy. By considering factors such as panel type, installation conditions, and financial options, beginners can navigate the complexities of solar energy systems and harness the power of the sun to meet their energy needs sustainably. As the world moves towards a more sustainable future, solar panel technology offers a practical and effective solution for reducing environmental impact and enhancing energy security.

Assessing Your Home's Solar Potential

Harnessing solar energy at home begins with a thorough assessment of your property's solar potential. This evaluation is crucial for determining whether solar panels are a viable option and how to maximize their efficiency. By understanding the key factors that influence solar potential, homeowners can make informed decisions and optimize their investment in renewable energy.

The first step in assessing solar potential is to evaluate the available sunlight. The amount of sunlight your home receives is influenced by geographic location, climate, and seasonal variations. Homes located in regions with high solar irradiance, such as the southwestern United States, are naturally more suited for solar energy. However, even in less sunny areas, solar panels can still be effective, provided they receive adequate sunlight throughout the year.

Roof orientation and tilt are critical factors in determining solar potential. Ideally, solar panels should face south in the northern hemisphere to capture the maximum amount of sunlight. East- or west-facing roofs can also be suitable, though they may produce slightly less energy. The angle of the roof should be close to the latitude of the location to optimize energy capture. Flat roofs offer flexibility, as panels can be mounted at the optimal angle using adjustable racks.

Shading is another important consideration. Trees, buildings, and other obstructions can cast shadows on solar panels, significantly reducing their efficiency. Conducting a shading analysis helps identify potential obstacles and determine the best placement for panels. Tools such as solar pathfinders or software applications can provide detailed insights into shading patterns throughout the year, allowing homeowners to make necessary adjustments.

The size and condition of the roof play a significant role in assessing solar potential. A structurally sound roof with ample space is essential for supporting solar panels. Before installation, it's advisable to inspect the roof for any damage or wear that might require repairs. Additionally, the age of the roof should be considered, as installing solar panels on a roof nearing the end of its lifespan may necessitate future removal and reinstallation.

Local regulations and zoning laws can impact the feasibility of solar installations. Some areas have restrictions on the placement or appearance of solar panels, while others offer incentives or streamlined permitting processes to encourage adoption. Researching local ordinances and consulting with professionals can help navigate these regulations and ensure compliance.

Financial considerations are integral to assessing solar potential. The initial cost of solar panels and installation can be substantial, but various financing options are available to make solar energy more accessible. Solar loans, leases, and power purchase agreements (PPAs) offer different pathways to ownership, allowing homeowners to choose the option that best suits their financial situation. Additionally, government incentives, rebates, and tax credits can significantly reduce the overall cost, making solar energy an attractive option for many.

Energy consumption patterns should also be evaluated when considering solar potential. Understanding your household's energy usage helps determine the size of the solar system needed to meet your needs. Reviewing past utility bills provides insights into average consumption and peak usage times. This information is crucial for designing a system that maximizes energy production and savings.

For those interested in energy independence, battery storage systems can complement solar panels by storing excess energy generated during the day for use at night or during power outages. While adding storage increases the initial investment, it enhances the reliability and flexibility of the solar system, providing greater control over energy usage.

Consulting with solar professionals is a valuable step in assessing your home's solar potential. Experienced installers can conduct site assessments, provide detailed proposals, and offer recommendations tailored to your specific needs. They can also assist with navigating the permitting process and ensuring that the installation complies with local codes and standards.

The environmental benefits of solar energy are significant, contributing to a reduction in greenhouse gas emissions and supporting efforts to combat climate change. By generating electricity on-site, homeowners can reduce their reliance on fossil fuels and contribute to a more sustainable future. Additionally, solar energy can enhance energy security by providing a reliable and renewable source of power.

As technology continues to advance, the efficiency and affordability of solar panels are expected to improve, further driving their adoption. Innovations in materials, manufacturing processes, and system integration are expanding the possibilities for solar energy, making it a key player in the transition to a clean energy future.

Assessing your home's solar potential empowers you to make informed decisions about adopting solar energy. By considering factors such as sunlight availability, roof orientation, shading, and financial options, you can navigate the complexities of solar energy systems and harness the power of the sun to meet your energy needs sustainably. As the world moves towards a more sustainable future, solar energy offers a practical and effective solution for reducing environmental impact and enhancing energy security.

Financial Considerations: Costs and Incentives

Navigating the financial landscape of renewable energy investments, particularly solar energy, requires a comprehensive understanding of both costs and incentives. This chapter provides a detailed exploration of the financial considerations involved in adopting solar technology, offering practical advice for beginners to make informed decisions.

The initial cost of solar panel installation can be a significant barrier for many homeowners. This cost typically includes the price of the solar panels themselves, inverters, mounting hardware, and installation labor. Additionally, there may be costs associated with permits, inspections, and any necessary upgrades to the electrical system. While the upfront investment can be substantial, it's important to consider the long-term savings and benefits that solar energy can provide.

One of the primary financial benefits of solar energy is the reduction in electricity bills. By generating your own electricity, you can significantly decrease your reliance on the grid, leading to lower monthly utility costs. The extent of these savings depends on factors such as the size of the solar system, local electricity rates, and your household's energy consumption patterns. In some cases, homeowners may even achieve net-zero energy consumption, where the solar system produces as much energy as the home uses over the course of a year.

To make solar energy more accessible, various financing options are available. Solar loans allow homeowners to finance the cost of the solar system over time, often with favorable interest rates. These loans can be secured or unsecured, with terms ranging from a few years to several decades. By spreading the cost over time, solar loans can make the investment more manageable and allow homeowners to start saving on electricity bills immediately.

Leasing is another option for those interested in solar energy without the upfront costs. With a solar lease, a third-party company installs and maintains the solar system on your property, and you pay a fixed monthly fee for the electricity generated. While leasing can provide immediate savings, it's important to carefully review the terms of the lease agreement, as it may include annual escalators or other conditions that affect long-term savings.

Power Purchase Agreements (PPAs) offer a similar arrangement to leasing, but instead of paying a fixed monthly fee, you pay for the electricity generated by the solar system at a predetermined rate. This rate is typically lower than the local utility rate, providing immediate savings. PPAs can be an attractive option for those who want to benefit from solar energy without the responsibility of ownership or maintenance.

Government incentives play a crucial role in reducing the overall cost of solar energy. The federal Investment Tax Credit (ITC) allows homeowners to deduct a percentage of the cost of installing a solar energy system from their federal taxes. This credit can significantly reduce the upfront cost and improve the return on investment. Additionally, many states and local governments offer rebates, tax credits, and other incentives to encourage the adoption of solar energy. These programs vary widely by location, so it's essential to research the specific incentives available in your area.

Net metering is another financial consideration that can enhance the value of solar energy. This billing arrangement allows homeowners to receive credit for excess electricity generated by their solar system and sent back to the grid. During times when the solar system produces more energy than the home consumes, the excess energy is fed into the grid, and the homeowner receives a credit on their utility bill. This credit can offset the cost of electricity drawn from the grid during periods of low solar production, such as at night or on cloudy days.

The payback period is an important metric for evaluating the financial viability of a solar investment. This period represents the time it takes for the savings from reduced electricity bills and incentives to equal the initial cost of the solar system. The payback period can vary depending on factors such as system size, local electricity rates, and available incentives. In many cases, homeowners can expect a payback period of five to ten years, after which the solar system continues to provide free electricity for the remainder of its lifespan.

For those interested in maximizing the financial benefits of solar energy, battery storage systems can be a valuable addition. By storing excess energy generated during the day, batteries provide a backup power source for use during peak demand periods or power outages. While adding storage increases the initial investment, it enhances the reliability and flexibility of the solar system, providing greater control over energy usage and potential savings.

It's important to consider the long-term financial implications of solar energy. Solar panels typically come with warranties of 20 to 25 years, and their lifespan can extend beyond 30 years with proper maintenance. Over this period, the savings on electricity bills can far exceed the initial investment, providing a substantial return on investment. Additionally, solar energy can increase the value of your home, making it more attractive to potential buyers.

Consulting with solar professionals and financial advisors can provide valuable insights into the costs and benefits of solar energy. These experts can help you navigate the complexities of financing options, incentives, and system design to ensure that your investment aligns with your financial goals.

As technology continues to advance and the cost of solar panels decreases, the financial case for solar energy becomes increasingly compelling. By understanding the costs, incentives, and financing options available, homeowners can make informed decisions and take advantage of the numerous benefits that solar energy offers. As the world moves towards a more sustainable future, solar energy provides a practical and effective solution for reducing environmental impact and enhancing energy security.

Choosing the Right Solar System for Beginners

Embarking on the journey to solar energy begins with selecting the right solar system tailored to your specific needs. For beginners, this process can seem daunting, but with a clear understanding of the key components and considerations, you can make an informed decision that maximizes both efficiency and savings.

The first step in choosing a solar system is to assess your energy needs. Understanding your household's energy consumption patterns is crucial for determining the size of the solar system required. Reviewing past utility bills provides insights into your average monthly usage and peak demand periods. This information helps in designing a system that meets your energy needs while optimizing cost-effectiveness.

Once you have a clear picture of your energy requirements, the next consideration is the type of solar panels. There are several options available, each with its own advantages. Monocrystalline panels, known for their high efficiency and sleek appearance, are ideal for installations with limited space. Polycrystalline panels offer a more cost-effective solution with slightly lower efficiency. Thin-film panels, which are lightweight and flexible, can be suitable for unconventional applications. The choice of panel type will depend on factors such as budget, available space, and aesthetic preferences.

Inverters are another critical component of a solar system. They convert the direct current (DC) generated by the solar panels into alternating current (AC), which is used by household appliances. There are three main types of inverters: string inverters, microinverters, and power optimizers. String inverters are the most common and cost-effective option, suitable for installations with minimal shading. Microinverters and power optimizers offer enhanced performance in shaded conditions by optimizing the output of individual panels, making them ideal for complex roof layouts.

The mounting system is an essential consideration for the installation of solar panels. Roof-mounted systems are the most common, utilizing existing roof space to support the panels. Ground-mounted systems offer flexibility in orientation and tilt, making them suitable for properties with ample land. Tracking systems, which adjust the panel's position to follow the sun's path, can further enhance energy capture but come at a higher cost.

Budget is a significant factor in choosing the right solar system. While the initial investment can be substantial, it's important to consider the long-term savings and benefits. Various financing options, such as solar loans, leases, and power purchase agreements (PPAs), can make solar energy more accessible. Solar loans allow you to finance the cost over time, while leases and PPAs provide immediate savings without the responsibility of ownership. Additionally, government incentives, rebates, and tax credits can significantly reduce the overall cost, making solar energy an attractive option for many.

The location and orientation of your property play a crucial role in the performance of a solar system. Ideally, solar panels should face south in the northern hemisphere to capture the maximum amount of sunlight. East- or west-facing roofs can also be suitable, though they may produce slightly less energy. The angle of the roof should be close to the latitude of the location to optimize energy capture. Conducting a shading analysis helps identify potential obstacles and determine the best placement for panels.

For those interested in energy independence, battery storage systems can complement solar panels by storing excess energy generated during the day for use at night or during power outages. While adding storage increases the initial investment, it enhances the reliability and flexibility of the solar system, providing greater control over energy usage.

Consulting with solar professionals is a valuable step in choosing the right solar system. Experienced installers can conduct site assessments, provide detailed proposals, and offer recommendations tailored to your specific needs. They can also assist with navigating the permitting process and ensuring that the installation complies with local codes and standards.

The environmental benefits of solar energy are significant, contributing to a reduction in greenhouse gas emissions and supporting efforts to combat climate change. By generating electricity on-site, homeowners can reduce their reliance on fossil fuels and contribute to a more sustainable future. Additionally, solar energy can enhance energy security by providing a reliable and renewable source of power.

As technology continues to advance, the efficiency and affordability of solar panels are expected to improve, further driving their adoption. Innovations in materials, manufacturing processes, and system integration are expanding the possibilities for solar energy, making it a key player in the transition to a clean energy future.

Choosing the right solar system empowers you to make informed decisions about adopting solar energy. By considering factors such as energy needs, panel type, inverter options, and financial considerations, you can navigate the complexities of solar energy systems and harness the power of the sun to meet your energy needs sustainably. As the world moves towards a more sustainable future, solar energy offers a practical and effective solution for reducing environmental impact and enhancing energy security.

Installation and Maintenance Essentials

Embarking on the journey of solar energy adoption involves not only selecting the right system but also understanding the intricacies of installation and maintenance. These elements are crucial for ensuring the longevity and efficiency of your solar investment. This chapter delves into the essential aspects of installing and maintaining solar panels, providing practical guidance for beginners.

The installation process begins with a comprehensive site assessment. This evaluation considers factors such as roof condition, available space, shading, and local regulations. A thorough assessment helps determine the optimal placement and orientation of solar panels, maximizing energy capture and efficiency. It's essential to work with experienced solar professionals who can conduct detailed site evaluations and provide tailored recommendations.

Once the site assessment is complete, the design and engineering phase commences. This stage involves selecting the appropriate solar panels, inverters, and mounting systems based on the specific characteristics of your property. The design should account for factors such as roof pitch, orientation, and potential shading to ensure optimal performance. Professional installers will create a detailed plan that outlines the layout and configuration of the solar system.

Permitting is a critical step in the installation process. Local regulations and zoning laws can impact the feasibility of solar installations, and obtaining the necessary permits is essential for compliance. The permitting process can vary widely depending on location, so it's important to research the specific requirements in your area. Solar professionals can assist with navigating the permitting process, ensuring that all necessary approvals are obtained before installation begins.

The actual installation of solar panels involves several key steps. First, the mounting system is securely attached to the roof or ground, providing a stable foundation for the panels. Next, the solar panels are carefully positioned and fastened to the mounting system, ensuring they are aligned for maximum sunlight exposure. The electrical components, including inverters and wiring, are then connected to integrate the solar system with the home's electrical system. Finally, the system undergoes a series of inspections and tests to verify its safety and functionality.

Safety is paramount during the installation process. Working at heights and handling electrical components requires specialized skills and equipment. Professional installers are trained to adhere to strict safety standards, minimizing the risk of accidents and ensuring a safe and reliable installation. Homeowners should avoid attempting DIY installations, as improper handling can lead to damage or injury.

Once the solar system is installed, regular maintenance is essential for ensuring its continued performance and longevity. Fortunately, solar panels are designed to be low-maintenance, with no moving parts and robust construction. However, periodic cleaning is necessary to remove dust, dirt, and debris that can accumulate on the surface of the panels and reduce their efficiency. In most cases, rain will naturally clean the panels, but in dry or dusty environments, manual cleaning may be required.

Inspections are another important aspect of solar maintenance. Regular inspections by qualified professionals can identify potential issues, such as loose connections, damaged components, or shading from new obstructions. Addressing these issues promptly can prevent more significant problems and ensure the system operates at peak efficiency. It's advisable to schedule inspections annually or biannually, depending on the specific conditions of your installation.

Monitoring the performance of your solar system is crucial for identifying any deviations from expected output. Many modern solar systems come equipped with monitoring software that provides real-time data on energy production and consumption. This information can help you track the system's performance, identify any anomalies, and make informed decisions about energy usage. If you notice a significant drop in energy production, it's important to investigate the cause and address any underlying issues.

In addition to routine maintenance, it's important to be aware of the warranty terms for your solar panels and components. Most solar panels come with warranties of 20 to 25 years, covering defects in materials and workmanship. Inverters and other components may have shorter warranty periods, so it's essential to understand the coverage and terms. Keeping detailed records of maintenance activities and inspections can be valuable if warranty claims are necessary.

The environmental benefits of solar energy extend beyond reducing greenhouse gas emissions. By generating electricity on-site, homeowners can reduce their reliance on fossil fuels and contribute to a more sustainable future. Additionally, solar energy can enhance energy security by providing a reliable and renewable source of power.

As technology continues to advance, the efficiency and affordability of solar panels are expected to improve, further driving their adoption. Innovations in materials, manufacturing processes, and system integration are expanding the possibilities for solar energy, making it a key player in the transition to a clean energy future.

Understanding the essentials of installation and maintenance empowers you to make informed decisions about adopting solar energy. By considering factors such as site assessment, design, permitting, and maintenance, you can navigate the complexities of solar energy systems and harness the power of the sun to meet your energy needs sustainably. As the world moves towards a more sustainable future, solar energy offers a practical and effective solution for reducing environmental impact and enhancing energy security.

Chapter 3: Exploring Wind Energy

How Wind Turbines Generate Electricity

Throughout history, women have played pivotal roles in shaping societies, cultures, and nations, often overcoming significant barriers to leave indelible marks on the world. Their stories are a testament to resilience, courage, and the relentless pursuit of justice and equality. By examining the lives and contributions of influential women, we gain insight into the diverse ways in which they have challenged norms, broken barriers, and inspired generations.

Cleopatra VII of Egypt, one of the most renowned figures of the ancient world, wielded power with intelligence and charisma. As the last active ruler of the Ptolemaic Kingdom of Egypt, she navigated the complex political landscape of the Roman Empire with strategic alliances and diplomatic acumen. Cleopatra's reign was marked by her efforts to restore Egypt's prosperity and independence, and her legacy endures as a symbol of female leadership and political savvy.

In the realm of science and mathematics, Hypatia of Alexandria stands out as a pioneering figure. Living in the 4th century CE, Hypatia was a philosopher, mathematician, and astronomer who led the Neoplatonic school in Alexandria. Her work in mathematics and her teachings on philosophy made her one of the most respected scholars of her time. Hypatia's tragic death at the hands of a mob underscores the challenges faced by women in intellectual pursuits, yet her legacy continues to inspire those who seek knowledge and truth.

The medieval period saw the rise of Joan of Arc, a peasant girl who became a national heroine of France. Claiming to have received visions from saints instructing her to support Charles VII and recover France from English domination, Joan led French forces to several important victories during the Hundred Years' War. Her unwavering faith and leadership in the face of adversity made her a symbol of courage and conviction. Despite her eventual capture and execution, Joan of Arc's legacy as a martyr and saint endures, inspiring countless individuals to stand up for their beliefs.

In the world of literature, Mary Wollstonecraft emerged as a trailblazer for women's rights. Her seminal work, "A Vindication of the Rights of Woman," published in 1792, argued for the education and empowerment of women, challenging the prevailing notion of female inferiority. Wollstonecraft's writings laid the groundwork for feminist thought and advocacy, influencing generations of women to fight for equality and justice.

The 19th century witnessed the remarkable achievements of Harriet Tubman, an abolitionist and political activist who escaped slavery and subsequently led hundreds of enslaved people to freedom via the Underground Railroad. Tubman's courage and determination in the face of immense danger made her a key figure in the abolitionist movement. Her efforts extended beyond the Underground Railroad, as she also served as a scout and spy for the Union Army during the Civil War. Tubman's legacy as a champion of freedom and equality continues to inspire those who fight against oppression.

In the realm of politics, the 20th century saw the rise of influential women leaders such as Indira Gandhi, the first and only female Prime Minister of India. Serving from 1966 to 1977 and again from 1980 until her assassination in 1984, Gandhi was a central figure in Indian politics. Her tenure was marked by significant economic and social reforms, as well as controversial decisions such as the declaration of a state of emergency. Despite the complexities of her leadership, Gandhi's impact on Indian politics and her role as a female leader in a male-dominated arena remain significant.

The fight for civil rights in the United States was profoundly shaped by the contributions of Rosa Parks, whose refusal to give up her seat on a segregated bus in Montgomery, Alabama, sparked the Montgomery Bus Boycott. Parks' act of defiance became a catalyst for the civil rights movement, highlighting the power of individual action in the struggle for justice. Her legacy as the "mother of the civil rights movement" underscores the importance of courage and resilience in the face of systemic oppression.

In the field of science, Marie Curie's groundbreaking research on radioactivity earned her two Nobel Prizes, making her the first woman to receive the prestigious award and the only person to win Nobel Prizes in two different scientific fields. Curie's dedication to scientific discovery and her contributions to the understanding of radioactivity have had a lasting impact on the fields of physics and chemistry. Her legacy as a pioneering scientist continues to inspire women in STEM fields to pursue their passions and break barriers.

The late 20th and early 21st centuries have seen the rise of influential women in various fields, from politics to entertainment. Figures such as Malala Yousafzai, the youngest Nobel Prize laureate, have become symbols of the fight for girls' education and empowerment. Malala's advocacy for education in the face of adversity has inspired a global movement for gender equality and access to education for all children.

In the arts, figures like Frida Kahlo have left an indelible mark on the world of visual art. Kahlo's unique style and exploration of identity, postcolonialism, and gender have made her an icon of artistic expression and feminist thought. Her work continues to resonate with audiences worldwide, challenging traditional notions of beauty and identity.

These stories of influential women throughout history highlight the diverse ways in which women have shaped the world. Their contributions, often achieved in the face of significant obstacles, serve as a testament to the power of resilience, courage, and determination. By recognizing and celebrating the achievements of these women, we honor their legacies and inspire future generations to continue the fight for equality and justice. Their stories remind us that progress is possible and that the pursuit of a more equitable world is a shared responsibility.

Small-Scale Wind Systems for Home Use

Harnessing the power of the wind has long been a pursuit of human ingenuity, and with the growing emphasis on sustainable energy, small-scale wind systems for home use have gained significant attention. These systems offer a promising solution for homeowners seeking to reduce their carbon footprint and achieve energy independence. Understanding the intricacies of small-scale wind systems, from their components to their installation and maintenance, is essential for anyone considering this renewable energy option.

At the heart of any wind energy system is the wind turbine, a device that converts the kinetic energy of the wind into electrical energy. Small-scale wind turbines, designed for residential use, typically have a capacity ranging from a few hundred watts to several kilowatts. These turbines can be mounted on rooftops or standalone towers, depending on the available space and wind conditions. The choice of turbine size and type is influenced by factors such as average wind speed, energy needs, and budget.

The efficiency of a wind turbine is largely determined by its location. Wind speed and consistency are critical factors in maximizing energy production. Ideally, a wind turbine should be situated in an area with unobstructed access to prevailing winds, away from buildings, trees, and other obstacles that can create turbulence and reduce efficiency. Conducting a thorough site assessment, including wind speed measurements and analysis of local wind patterns, is a crucial step in the planning process.

In addition to the turbine, a small-scale wind system comprises several key components, including a tower, a controller, an inverter, and a battery storage system. The tower elevates the turbine to capture stronger and more consistent winds, while the controller regulates the flow of electricity from the turbine to the inverter and battery. The inverter converts the direct current (DC) generated by the turbine into alternating current (AC), which can be used to power household appliances. A battery storage system allows homeowners to store excess energy for use during periods of low wind or high demand.

The installation of a small-scale wind system requires careful planning and consideration of various factors, including zoning regulations, permitting requirements, and potential environmental impacts. Local zoning laws may dictate the allowable height and placement of wind turbines, and obtaining the necessary permits can be a complex process. Additionally, it's important to consider the potential impact on local wildlife, particularly birds and bats, and to implement measures to mitigate any negative effects.

Once installed, a small-scale wind system requires regular maintenance to ensure optimal performance and longevity. Routine inspections of the turbine, tower, and electrical components are essential to identify and address any issues before they become significant problems. This includes checking for wear and tear on moving parts, ensuring that electrical connections are secure, and monitoring the performance of the battery storage system. Regular maintenance not only extends the lifespan of the system but also maximizes energy production and efficiency.

The financial considerations of small-scale wind systems are an important aspect for homeowners to evaluate. While the initial investment can be substantial, the long-term savings on energy bills and potential incentives, such as tax credits and rebates, can offset the costs. Additionally, the value of energy independence and the environmental benefits of reducing reliance on fossil fuels are significant factors to consider. Conducting a cost-benefit analysis, taking into account the expected energy production, installation costs, and available incentives, can help homeowners make informed decisions.

For those interested in small-scale wind systems, engaging with experienced professionals and industry experts can provide valuable insights and guidance. From site assessment and system design to installation and maintenance, working with knowledgeable professionals ensures that the system is tailored to meet the specific needs and conditions of the home. Additionally, connecting with other homeowners who have installed wind systems can offer practical advice and firsthand experiences.

The potential of small-scale wind systems extends beyond individual homes, contributing to broader efforts to promote renewable energy and sustainability. By generating clean, renewable energy, these systems help reduce greenhouse gas emissions and decrease reliance on non-renewable energy sources. As more homeowners adopt wind energy, the collective impact on the environment and energy landscape can be substantial.

In the context of a rapidly changing energy landscape, small-scale wind systems represent a viable and sustainable option for homeowners seeking to embrace renewable energy. By understanding the components, installation process, and maintenance requirements, individuals can make informed decisions and take meaningful steps towards energy independence. The journey to harnessing wind energy is not without its challenges, but the rewards of contributing to a more sustainable future are well worth the effort.

The insights gained from exploring small-scale wind systems for home use offer valuable perspectives on the potential for growth and transformation in the renewable energy sector. By recognizing the opportunities and challenges associated with wind energy, homeowners can navigate the complexities of this technology with greater awareness and intention. These insights remind us of the importance of ethical and responsible use of energy resources, as well as the potential for empowerment and positive change.

In conclusion, small-scale wind systems for home use provide a practical and sustainable solution for homeowners seeking to reduce their environmental impact and achieve energy independence. By understanding the intricacies of wind energy systems and engaging with experienced professionals, individuals can make informed decisions and contribute to a more sustainable energy future. Recognizing and valuing the potential of renewable energy is essential for fostering innovation, creativity, and positive change in the energy landscape and beyond.

Evaluating Wind Potential in Your Area

In the intricate tapestry of workplace dynamics, gender plays a pivotal role, influencing interactions, opportunities, and outcomes. Examining real-world case studies of gender dynamics at work provides valuable insights into the challenges and successes experienced by individuals and organizations striving for equality. These stories illuminate the complexities of gender in the workplace and offer lessons that can guide efforts to create more inclusive and equitable environments.

One notable case involves a multinational technology company that embarked on a journey to address gender disparities within its workforce. Despite being a leader in innovation, the company faced criticism for its lack of gender diversity, particularly in technical and leadership roles. In response, the company launched a comprehensive initiative aimed at increasing the representation of women across all levels. This initiative included targeted recruitment efforts, mentorship programs, and leadership development opportunities for women. Over time, the company saw a significant increase in the number of women in technical roles and leadership positions, demonstrating the impact of intentional and sustained efforts to promote gender diversity.

Another compelling case study highlights the experiences of a female executive in the finance industry, a sector traditionally dominated by men. Despite her qualifications and achievements, she encountered numerous barriers to advancement, including biased assumptions about her capabilities and a lack of access to influential networks. Determined to succeed, she leveraged her expertise and built a strong support network of mentors and allies. Her perseverance paid off, as she eventually rose to a senior leadership position, where she became a vocal advocate for gender equality and mentorship for other women in the industry. Her story underscores the importance of resilience and the power of mentorship in overcoming gender-based obstacles.

In contrast, a case study from the healthcare sector reveals the challenges faced by male nurses in a predominantly female profession. Despite the growing demand for nurses, men in the field often encounter stereotypes and biases that question their suitability for caregiving roles. One male nurse shared his experiences of being perceived as less compassionate or nurturing than his female colleagues, despite his dedication and competence. To address these challenges, he became involved in initiatives to promote diversity and inclusion within the nursing profession, advocating for the recognition of diverse skills and perspectives. His efforts contributed to a more inclusive environment where all nurses, regardless of gender, were valued for their contributions.

A case from the creative industry illustrates the impact of gender dynamics on collaboration and creativity. A mixed-gender team working on a high-profile advertising campaign experienced tension and conflict due to differing communication styles and expectations. The team leader, recognizing the potential for gender dynamics to influence team interactions, facilitated open discussions about communication preferences and encouraged team members to share their perspectives. By fostering an environment of mutual respect and understanding, the team was able to harness their diverse strengths and produce a successful campaign that resonated with a wide audience. This case highlights the importance of addressing gender dynamics in collaborative settings to enhance creativity and innovation.

In the realm of academia, a case study explores the experiences of women in STEM (science, technology, engineering, and mathematics) fields. Despite efforts to increase female representation, women in STEM often face challenges such as gender bias, lack of mentorship, and work-life balance issues. One university implemented a program to support female faculty and students in STEM, offering mentorship, networking opportunities, and resources for career development. The program led to increased retention and advancement of women in STEM, demonstrating the effectiveness of targeted support and community-building efforts.

These case studies reveal common themes and lessons that can inform efforts to address gender dynamics in the workplace. First, intentional and sustained efforts are essential for promoting gender diversity and inclusion. Organizations must commit to creating equitable opportunities and addressing systemic barriers that hinder progress. Second, mentorship and support networks play a crucial role in empowering individuals to overcome challenges and achieve their goals. Providing access to mentors and allies can help individuals navigate complex dynamics and build confidence in their abilities.

Additionally, fostering open communication and understanding is key to addressing gender dynamics and enhancing collaboration. Encouraging dialogue about diverse perspectives and experiences can lead to more inclusive and innovative outcomes. Finally, recognizing and valuing diverse skills and contributions is essential for creating environments where all individuals feel respected and valued.

The insights gained from these case studies offer valuable perspectives on the complexities of gender dynamics at work and the potential for growth and transformation. By understanding the factors that influence gender dynamics and implementing strategies to address them, organizations can create more inclusive and equitable environments. These efforts remind us of the importance of ethical and responsible use of power, as well as the potential for empowerment and positive change.

In conclusion, examining case studies of gender dynamics at work provides valuable insights into the challenges and successes experienced by individuals and organizations striving for equality. By learning from these stories and implementing strategies to promote diversity and inclusion, we can work towards creating environments where all individuals have the opportunity to thrive and contribute to the collective well-being of society. Recognizing and valuing diverse experiences and perspectives is essential for fostering innovation, creativity, and positive change in the workplace and beyond.

Benefits and Challenges of Wind Power

Media wields an undeniable influence over societal norms, shaping perceptions and expectations in ways both overt and subtle. Among the most profound impacts is the media's role in defining and perpetuating gender norms. These norms, which dictate the behaviors, roles, and attributes deemed appropriate for men and women, are often reinforced through various forms of media, including television, film, advertising, and digital platforms. Understanding how media influences gender norms is crucial for fostering a more equitable and inclusive society.

From a young age, individuals are exposed to media portrayals that shape their understanding of gender roles. Children's television shows and movies often feature characters that embody traditional gender stereotypes: boys are adventurous and strong, while girls are nurturing and passive. These early impressions can have a lasting impact, influencing children's self-perception and aspirations. For instance, a young girl who sees female characters primarily in caregiving roles may internalize the belief that her value lies in nurturing others, while a boy who sees male characters as heroes may feel pressured to conform to ideals of strength and bravery.

As individuals grow older, the media continues to play a significant role in reinforcing gender norms. Advertisements, in particular, are powerful vehicles for conveying messages about gender. They often depict men and women in stereotypical roles, with men portrayed as assertive and career-focused, and women as homemakers or objects of beauty. These portrayals not only reinforce existing stereotypes but also create unrealistic standards that individuals may feel compelled to meet. The pressure to conform to these ideals can lead to feelings of inadequacy and dissatisfaction, particularly when individuals do not see themselves reflected in the media they consume.

The impact of media on gender norms extends beyond individual perceptions to influence broader societal attitudes and behaviors. Media representations can shape public discourse and policy by framing certain issues in specific ways. For example, the portrayal of women in leadership roles in media can challenge traditional gender norms and inspire real-world change by normalizing the idea of women in positions of power. Conversely, the underrepresentation of women and minorities in media can perpetuate the status quo, reinforcing the notion that certain roles or industries are not accessible to all.

In recent years, there has been a growing recognition of the need for more diverse and accurate representations of gender in media. This shift is driven by advocacy from audiences, creators, and organizations who understand the power of media to shape cultural narratives. As a result, there has been an increase in media content that challenges traditional gender norms and offers more nuanced portrayals of individuals. Films and television shows are beginning to feature strong, complex female characters who defy stereotypes, as well as male characters who embrace vulnerability and emotional depth.

The rise of digital media and social platforms has also played a significant role in diversifying gender representation. These platforms provide a space for marginalized voices to share their stories and perspectives, challenging mainstream narratives and offering alternative representations of gender. Social media campaigns and movements, such as #MeToo and #TimesUp, have brought attention to issues of gender inequality and harassment, prompting conversations and actions that challenge the status quo.

Despite these positive developments, challenges remain in achieving truly equitable gender representation in media. The industry continues to grapple with issues of diversity and inclusion, both in front of and behind the camera. Women, particularly women of color, LGBTQ+ individuals, and those with disabilities, remain underrepresented in key creative and decision-making roles. This lack of diversity can result in media content that fails to accurately reflect the experiences and identities of all individuals.

To address these challenges, it is essential for media creators and organizations to prioritize diversity and inclusion in their work. This includes actively seeking out and amplifying underrepresented voices, both in storytelling and in leadership positions. By fostering an environment where diverse perspectives are valued and celebrated, the media industry can create content that resonates with a broader audience and reflects the richness of human experience.

Media literacy is another important tool for addressing gender representation. By equipping audiences with the skills to critically analyze media content, individuals can become more aware of the ways in which gender is portrayed and the impact of these portrayals. Media literacy education can empower individuals to question stereotypes, recognize bias, and advocate for more inclusive and accurate representations.

The role of consumers in shaping media representation should not be underestimated. Audiences have the power to influence the media landscape through their choices and voices. By supporting content that challenges traditional gender norms and promotes diversity, consumers can send a message to creators and organizations about the types of stories and representations they value. Engaging in conversations and advocacy around media representation can also contribute to a cultural shift towards greater inclusivity and equity.

The insights gained from examining media influence on gender norms offer valuable perspectives on the complexities of cultural narratives and the potential for growth and transformation. By understanding the factors that influence media portrayals and implementing strategies to address them, we can work towards creating a media landscape that reflects and celebrates the diversity of human experience. These efforts remind us of the importance of ethical and responsible storytelling, as well as the potential for empowerment and positive change.

In conclusion, media plays a powerful role in shaping gender norms and influencing societal perceptions. By examining the ways in which gender is portrayed and advocating for more diverse and accurate representations, we can work towards creating a media landscape that reflects the richness of human experience and promotes equity and inclusion. Recognizing and valuing diverse stories and perspectives is essential for fostering innovation, creativity, and positive change in the media industry and beyond.

Case Studies: Successful Wind Energy Projects

The story of wind energy is one of innovation, perseverance, and transformation. Across the globe, numerous wind energy projects have emerged as beacons of success, demonstrating the potential of harnessing the wind to power our world sustainably.

These projects not only provide valuable lessons in engineering and environmental stewardship but also highlight the importance of community engagement, policy support, and technological advancement. By examining a selection of successful wind energy projects, we can gain insights into the factors that contribute to their success and the challenges they have overcome.

One of the most notable examples of a successful wind energy project is the Horns Rev 2 offshore wind farm in Denmark. Located in the North Sea, this project was completed in 2009 and was, at the time, the largest offshore wind farm in the world. With 91 turbines and a total capacity of 209 megawatts, Horns Rev 2 generates enough electricity to power approximately 200,000 Danish households. The project's success can be attributed to Denmark's strong commitment to renewable energy, supported by favorable government policies and incentives. The Danish government's long-term energy strategy, which includes ambitious targets for wind energy, has created a stable and supportive environment for investment and development.

Horns Rev 2 also exemplifies the importance of technological innovation in wind energy projects. The use of advanced turbine technology, capable of withstanding harsh offshore conditions, has been crucial to the project's success. Additionally, the integration of cutting-edge monitoring and maintenance systems ensures the efficient operation of the wind farm, minimizing downtime and maximizing energy production. This focus on technology and innovation has positioned Denmark as a global leader in wind energy, inspiring other countries to follow suit.

Another exemplary wind energy project is the Alta Wind Energy Center in California, USA. As one of the largest onshore wind farms in the world, Alta boasts a capacity of over 1,500 megawatts, providing clean energy to hundreds of thousands of homes. The project's success is rooted in its strategic location in the Tehachapi Pass, an area known for its strong and consistent winds. This natural advantage, combined with California's progressive renewable energy policies, has made Alta a cornerstone of the state's clean energy portfolio.

The development of the Alta Wind Energy Center also underscores the importance of collaboration between public and private sectors. The project was made possible through partnerships between energy companies, government agencies, and local communities. These collaborations facilitated the necessary permitting and regulatory approvals, as well as the construction of transmission infrastructure to deliver the generated electricity to consumers. By working together, stakeholders were able to overcome challenges and ensure the project's success.

In Australia, the Macarthur Wind Farm stands as a testament to the potential of large-scale wind energy projects. Located in Victoria, Macarthur is the largest wind farm in the Southern Hemisphere, with a capacity of 420 megawatts. The project was developed through a joint venture between two major energy companies, highlighting the role of strategic partnerships in advancing wind energy. Macarthur's success is also attributed to its comprehensive environmental impact assessments and community engagement efforts, which addressed concerns and garnered local support.

The Macarthur Wind Farm's development process involved extensive consultation with local communities and stakeholders. By engaging with residents and addressing their concerns, the project developers were able to build trust and foster a sense of ownership among the community. This approach not only facilitated the project's approval but also ensured its long-term success by creating a positive relationship between the wind farm and the surrounding area.

In China, the Gansu Wind Farm Project represents a monumental achievement in wind energy development. As part of China's ambitious plan to increase its renewable energy capacity, the Gansu Wind Farm is one of the largest wind power bases in the world, with a planned capacity of 20,000 megawatts. The project's success is driven by China's commitment to reducing its reliance on coal and addressing environmental concerns. Government support, in the form of subsidies and favorable policies, has been instrumental in attracting investment and driving the rapid expansion of wind energy in the region.

The Gansu Wind Farm Project also highlights the importance of infrastructure development in supporting large-scale wind energy projects. The construction of new transmission lines and grid upgrades has been essential to accommodate the increased capacity and ensure the efficient distribution of electricity. This focus on infrastructure demonstrates the need for a holistic approach to wind energy development, where generation, transmission, and distribution are all considered in tandem.

In the United Kingdom, the Whitelee Wind Farm serves as a shining example of successful wind energy integration into the national grid. Located near Glasgow, Scotland, Whitelee is the largest onshore wind farm in the UK, with a capacity of 539 megawatts. The project's success is attributed to its strategic location, strong government support, and commitment to environmental stewardship. Whitelee's development involved extensive environmental assessments and habitat restoration efforts, ensuring that the project minimized its impact on local ecosystems.

Whitelee Wind Farm also emphasizes the importance of public engagement and education in wind energy projects. The site features a visitor center that provides information about renewable energy and the benefits of wind power. By educating the public and promoting awareness, Whitelee has fostered a positive perception of wind energy and encouraged broader acceptance of renewable technologies.

These case studies of successful wind energy projects offer valuable lessons for future developments. Key factors contributing to their success include strong government support, strategic partnerships, technological innovation, and community engagement. By prioritizing these elements, wind energy projects can overcome challenges and achieve their full potential, contributing to a more sustainable and resilient energy future.

The insights gained from these projects remind us of the importance of ethical and responsible energy development, as well as the potential for empowerment and positive change. By recognizing and valuing the potential of renewable energy, we can foster innovation, creativity, and positive change in the energy landscape and beyond.

Chapter 4: Hydropower and Its Applications

Understanding Hydroelectric Power Generation

Hydroelectric power generation stands as one of the most established and reliable forms of renewable energy, harnessing the natural flow of water to produce electricity. This method of power generation has been utilized for over a century, providing a significant portion of the world's electricity supply. Understanding the intricacies of hydroelectric power involves exploring its mechanisms, benefits, challenges, and future potential.

At its core, hydroelectric power generation relies on the conversion of kinetic energy from flowing or falling water into mechanical energy, which is then transformed into electrical energy. This process typically involves a dam constructed across a river, creating a reservoir or a height difference that allows water to flow through turbines. As water passes through these turbines, it spins them, driving generators that produce electricity. The amount of electricity generated depends on the volume of water flow and the height from which the water falls, known as the head.

One of the primary advantages of hydroelectric power is its ability to provide a consistent and reliable source of electricity. Unlike solar or wind energy, which are subject to weather conditions, hydroelectric plants can operate continuously, providing a stable base load of power. This reliability makes hydroelectric power an attractive option for meeting the energy demands of large populations and industries.

Hydroelectric power is also a clean and sustainable energy source, producing no direct emissions of greenhouse gases or air pollutants. This environmental benefit is particularly significant in the context of global efforts to reduce carbon emissions and combat climate change. By replacing fossil fuel-based power generation with hydroelectricity, countries can significantly decrease their carbon footprint and contribute to a more sustainable energy future.

In addition to its environmental benefits, hydroelectric power offers economic advantages. Once a hydroelectric plant is constructed, its operational and maintenance costs are relatively low compared to fossil fuel plants. The longevity of hydroelectric facilities, which can operate for several decades, further enhances their economic viability. Moreover, hydroelectric projects can stimulate local economies by creating jobs during the construction and operation phases and by providing a reliable source of electricity that supports industrial and economic development.

Despite these advantages, hydroelectric power generation is not without its challenges. The construction of dams and reservoirs can have significant environmental and social impacts, including the displacement of communities, alteration of ecosystems, and changes in water quality and flow patterns. These impacts necessitate careful planning and management to minimize negative consequences and ensure the sustainability of hydroelectric projects.

The environmental impact of hydroelectric power is particularly pronounced in terms of its effects on aquatic ecosystems. Dams can disrupt the natural migration patterns of fish and other aquatic species, leading to declines in biodiversity and changes in ecosystem dynamics. To address these challenges, many hydroelectric projects incorporate fish ladders, bypass channels, and other measures to facilitate the movement of aquatic species and mitigate ecological impacts.

Social considerations are also critical in the development of hydroelectric projects. The displacement of communities and changes in land use can have profound effects on local populations, necessitating comprehensive resettlement and compensation plans. Engaging with affected communities and stakeholders throughout the planning and implementation process is essential to ensure that their needs and concerns are addressed and that the benefits of the project are equitably distributed.

Technological advancements are playing a crucial role in addressing the challenges associated with hydroelectric power generation. Innovations in turbine design, for example, are improving the efficiency and environmental performance of hydroelectric plants. Variable-speed turbines and advanced control systems allow for more precise management of water flow and energy output, optimizing the performance of hydroelectric facilities and reducing their environmental impact.

The integration of hydroelectric power with other renewable energy sources is another promising trend. By combining hydroelectricity with solar, wind, or other renewable technologies, energy systems can achieve greater flexibility and resilience. For example, pumped-storage hydroelectricity, which involves pumping water to a higher elevation during periods of low electricity demand and releasing it to generate power during peak demand, can complement intermittent renewable sources and provide grid stability.

The future of hydroelectric power generation is also being shaped by the exploration of new and innovative approaches. Small-scale and micro-hydroelectric projects, which have a lower environmental impact and can be deployed in remote or off-grid locations, are gaining attention as a means of expanding access to clean energy. These projects can provide electricity to rural communities, support local development, and contribute to energy security.

The potential for retrofitting existing infrastructure, such as non-powered dams and water supply systems, with hydroelectric capabilities is another avenue for expanding hydroelectric power generation. By leveraging existing structures, these projects can minimize environmental and social impacts while increasing renewable energy capacity.

Understanding hydroelectric power generation requires a comprehensive examination of its mechanisms, benefits, challenges, and future potential. By recognizing the complexities and opportunities associated with this form of renewable energy, stakeholders can make informed decisions that maximize its benefits while addressing its challenges. The insights gained from this exploration remind us of the importance of ethical and responsible energy development, as well as the potential for empowerment and positive change.

The continued advancement of hydroelectric power generation holds promise for a more sustainable and resilient energy future. By embracing innovation, collaboration, and sustainability, hydroelectric power can play a vital role in meeting the world's energy needs while preserving the environment and supporting communities. Recognizing and valuing the potential of hydroelectric power is essential for fostering innovation, creativity, and positive change in the energy landscape and beyond.

Micro-Hydropower Systems for Residential Use

Micro-hydropower systems offer an innovative and sustainable solution for residential energy needs, harnessing the power of flowing water to generate electricity on a small scale. These systems are particularly appealing for homeowners with access to a reliable water source, such as a stream or river, providing an opportunity to reduce reliance on traditional energy sources and lower utility bills. Understanding the components, benefits, and considerations of micro-hydropower systems is essential for those interested in exploring this renewable energy option.

At the heart of a micro-hydropower system is the conversion of water's kinetic energy into electrical energy. This process begins with the diversion of water from a natural source through a channel or pipeline, known as a penstock, which directs the flow towards a turbine. As water flows through the turbine, it spins the blades, driving a generator that produces electricity. The amount of electricity generated depends on the flow rate and the head, or the vertical distance the water falls. These factors determine the potential energy available for conversion.

One of the primary advantages of micro-hydropower systems is their ability to provide a consistent and reliable source of electricity. Unlike solar or wind energy, which are subject to fluctuations in weather conditions, micro-hydropower can operate continuously as long as there is a steady flow of water. This reliability makes it an attractive option for off-grid homes or those seeking to supplement their existing energy supply with renewable sources.

Micro-hydropower systems are also environmentally friendly, producing no direct emissions or pollutants. By utilizing the natural flow of water, these systems have a minimal impact on the environment compared to larger hydroelectric projects that require damming and significant alteration of ecosystems. Additionally, micro-hydropower systems can contribute to energy independence and resilience, reducing reliance on fossil fuels and enhancing the sustainability of residential energy consumption.

For homeowners considering a micro-hydropower system, several key factors must be taken into account. The first consideration is the availability and reliability of a suitable water source. A consistent flow of water is essential for the system's operation, and seasonal variations in water levels can affect energy production. Conducting a thorough assessment of the water source, including flow rate measurements and seasonal patterns, is crucial to determine the feasibility of a micro-hydropower system.

The site location and topography also play a significant role in the design and installation of a micro-hydropower system. The head, or the vertical drop from the water source to the turbine, is a critical factor in determining the system's efficiency and energy output. Sites with a higher head can generate more electricity with a smaller flow rate, making them ideal for micro-hydropower systems. Additionally, the proximity of the water source to the residence and the ease of access for installation and maintenance are important considerations.

The selection of appropriate equipment is another essential aspect of micro-hydropower system design. Turbines come in various types, each suited to different flow rates and heads. For example, Pelton turbines are ideal for high-head, low-flow sites, while Kaplan turbines are better suited for low-head, high-flow conditions. Choosing the right turbine and generator combination is crucial to maximizing the system's efficiency and energy output.

Permitting and regulatory requirements must also be addressed when planning a micro-hydropower system. Depending on the location and scale of the project, permits may be required from local, state, or federal agencies. These permits ensure that the system complies with environmental regulations and does not adversely affect water quality or aquatic habitats. Engaging with regulatory authorities early in the planning process can help streamline the permitting process and address any potential concerns.

The cost of installing a micro-hydropower system can vary widely depending on factors such as site conditions, equipment selection, and installation complexity. While the initial investment may be significant, the long-term savings on energy bills and the potential for government incentives or rebates can offset these costs. Conducting a cost-benefit analysis and exploring financing options can help homeowners make informed decisions about the feasibility and affordability of a micro-hydropower system.

Maintenance and monitoring are essential to ensure the continued performance and longevity of a micro-hydropower system. Regular inspections of the turbine, generator, and water intake system can help identify and address any issues before they become significant problems. Monitoring the system's energy output and water flow can also provide valuable data for optimizing performance and making adjustments as needed.

The integration of micro-hydropower systems with other renewable energy sources, such as solar or wind, can enhance the overall sustainability and resilience of residential energy systems. By combining different energy sources, homeowners can create a hybrid system that maximizes energy production and reliability, even in varying weather conditions. This approach can also provide greater flexibility in meeting energy demands and reducing reliance on grid electricity.

Micro-hydropower systems offer a promising and sustainable solution for residential energy needs, harnessing the power of water to generate clean electricity. By understanding the components, benefits, and considerations of these systems, homeowners can make informed decisions about their potential for energy independence and sustainability. The insights gained from exploring micro-hydropower systems remind us of the importance of innovation and responsible energy development, as well as the potential for empowerment and positive change.

The continued advancement of micro-hydropower technology holds promise for a more sustainable and resilient energy future. By embracing innovation, collaboration, and sustainability, micro-hydropower systems can play a vital role in meeting residential energy needs while preserving the environment and supporting communities. Recognizing and valuing the potential of micro-hydropower is essential for fostering innovation, creativity, and positive change in the energy landscape and beyond.

Environmental Considerations and Impact

The relationship between energy production and the environment is a complex and multifaceted one, with significant implications for ecosystems, biodiversity, and human health. As the world increasingly turns to renewable energy sources to meet its growing energy demands, understanding the environmental considerations and impacts of these technologies is crucial. This chapter delves into the environmental aspects of energy production, focusing on the balance between harnessing renewable resources and preserving the natural world.

Renewable energy sources, such as wind, solar, and hydroelectric power, are often lauded for their minimal carbon emissions and potential to mitigate climate change. However, each of these technologies carries its own set of environmental considerations that must be addressed to ensure sustainable development. For instance, wind energy, while clean and abundant, can pose risks to avian and bat populations. The placement of wind turbines in migratory paths or near habitats can lead to collisions and habitat disruption. To mitigate these impacts, careful site selection and the implementation of technologies such as radar systems to detect and deter wildlife are essential.

Solar energy, another cornerstone of the renewable revolution, presents its own environmental challenges. The production of photovoltaic panels involves the use of hazardous materials and energy-intensive processes, which can contribute to pollution and resource depletion. Additionally, large-scale solar farms require significant land use, potentially leading to habitat loss and fragmentation. To address these concerns, the industry is moving towards more sustainable manufacturing practices, recycling programs for solar panels, and the development of solar installations on already disturbed lands, such as rooftops and brownfields.

Hydroelectric power, one of the oldest forms of renewable energy, is often associated with significant environmental impacts due to the construction of dams and reservoirs. These structures can alter river ecosystems, disrupt fish migration, and affect water quality. The inundation of land for reservoirs can also lead to the displacement of communities and wildlife. To minimize these impacts, modern hydroelectric projects are increasingly incorporating fish ladders, bypass channels, and other measures to facilitate the movement of aquatic species. Additionally, the development of small-scale and run-of-river hydroelectric systems offers a less intrusive alternative, harnessing the power of flowing water without the need for large dams.

Beyond the specific impacts of individual renewable technologies, the broader environmental considerations of energy production include land use, resource consumption, and waste generation. The transition to renewable energy requires careful planning and management to balance the need for clean energy with the preservation of natural landscapes and ecosystems. This involves not only the selection of appropriate sites for energy projects but also the integration of energy production with other land uses, such as agriculture, conservation, and recreation.

Resource consumption is another critical consideration in the environmental impact of energy production. The extraction and processing of materials for renewable energy technologies, such as rare earth metals for wind turbines and lithium for batteries, can have significant environmental and social consequences. Sustainable sourcing practices, recycling, and the development of alternative materials are essential strategies for reducing the resource footprint of renewable energy.

Waste generation, particularly in the form of decommissioned equipment and materials, is an emerging concern for the renewable energy sector. As the first generation of renewable energy installations reaches the end of its lifecycle, the industry faces the challenge of managing and recycling large volumes of waste. Innovative recycling technologies and circular economy approaches are being developed to address this issue, ensuring that materials are recovered and reused rather than ending up in landfills.

The environmental considerations of energy production extend beyond the immediate impacts of individual projects to encompass broader ecological and social dimensions. The transition to renewable energy presents an opportunity to rethink our relationship with the environment and to develop energy systems that are not only sustainable but also equitable and just. This involves engaging with local communities, respecting indigenous rights, and ensuring that the benefits of renewable energy are shared fairly.

Public participation and stakeholder engagement are critical components of environmentally responsible energy development. By involving communities in the planning and decision-making processes, energy projects can better address local concerns and priorities, leading to more sustainable and accepted outcomes. Transparent communication and collaboration with stakeholders, including government agencies, non-governmental organizations, and industry, are essential for building trust and fostering a shared vision for a sustainable energy future.

The integration of environmental considerations into energy policy and planning is another key aspect of sustainable energy development. Policymakers play a crucial role in setting the framework for environmentally responsible energy production, through regulations, incentives, and support for research and innovation. By prioritizing environmental sustainability in energy policy, governments can drive the transition to renewable energy while safeguarding ecosystems and communities.

The environmental considerations and impacts of energy production are complex and multifaceted, requiring a holistic and integrated approach to address them effectively. By understanding and addressing these considerations, the renewable energy sector can contribute to a more sustainable and resilient energy future. The insights gained from exploring the environmental dimensions of energy production remind us of the importance of ethical and responsible energy development, as well as the potential for empowerment and positive change.

The continued advancement of renewable energy technologies and practices holds promise for a more sustainable and resilient energy future. By embracing innovation, collaboration, and sustainability, the energy sector can play a vital role in meeting global energy needs while preserving the environment and supporting communities. Recognizing and valuing the potential of renewable energy is essential for fostering innovation, creativity, and positive change in the energy landscape and beyond.

Innovations in Hydropower Technology

Hydropower has long been a cornerstone of renewable energy, providing a reliable and sustainable source of electricity. However, as the world seeks to expand its renewable energy portfolio and address environmental concerns, innovations in hydropower technology are becoming increasingly important. These advancements are not only enhancing the efficiency and sustainability of hydropower systems but also opening up new possibilities for harnessing the power of water in innovative ways.

One of the most significant innovations in hydropower technology is the development of advanced turbine designs. Traditional turbines, while effective, often face limitations in terms of efficiency and environmental impact. New turbine designs, such as the fish-friendly turbines, are addressing these challenges by minimizing harm to aquatic life and improving energy conversion efficiency. These turbines feature specially designed blades and flow patterns that allow fish to pass through safely, reducing the ecological impact of hydropower plants.

Variable-speed turbines represent another breakthrough in hydropower technology. Unlike conventional turbines that operate at a fixed speed, variable-speed turbines can adjust their rotational speed to match the flow of water. This flexibility allows for more efficient energy capture, particularly in sites with fluctuating water levels or flow rates. By optimizing the turbine's performance in real-time, variable-speed technology enhances the overall efficiency and output of hydropower systems.

The integration of digital technologies and data analytics is also transforming the hydropower sector. Smart hydropower systems leverage sensors, data analytics, and machine learning algorithms to monitor and optimize plant operations. These systems can predict maintenance needs, optimize water flow, and enhance grid integration, leading to improved efficiency and reduced operational costs. By harnessing the power of data, hydropower operators can make informed decisions that maximize energy production and minimize environmental impact.

Pumped-storage hydropower is another area of innovation that is gaining attention. This technology involves storing energy by pumping water to a higher elevation during periods of low electricity demand and releasing it to generate power during peak demand. Recent advancements in pumped-storage technology are focusing on improving efficiency and reducing environmental impact. For example, closed-loop systems, which use reservoirs that are not connected to natural water bodies, minimize ecological disruption and offer greater flexibility in site selection.

The exploration of low-head and small-scale hydropower systems is expanding the potential for hydropower in regions where traditional large-scale projects are not feasible. Low-head hydropower systems can operate in sites with minimal elevation differences, making them suitable for rivers and canals with gentle slopes. These systems often utilize innovative turbine designs, such as Archimedes screws or hydrokinetic turbines, to capture energy from slow-moving water. Small-scale hydropower projects, including micro and pico hydropower systems, are providing renewable energy solutions for remote and off-grid communities, enhancing energy access and resilience.

Floating hydropower technology is an emerging innovation that offers new possibilities for energy generation. By placing turbines on floating platforms, this technology can harness the energy of water bodies without the need for dams or significant infrastructure. Floating hydropower systems are particularly well-suited for reservoirs, lakes, and coastal areas, where they can complement existing energy infrastructure and provide additional capacity. The modular nature of floating systems allows for easy scalability and adaptability to different site conditions.

The integration of hydropower with other renewable energy sources is another promising trend. Hybrid systems that combine hydropower with solar, wind, or biomass energy can enhance the overall reliability and flexibility of renewable energy systems. For example, solar panels can be installed on the surface of reservoirs, providing additional energy generation while reducing evaporation and preserving water resources. By leveraging the complementary strengths of different energy sources, hybrid systems can provide a more stable and resilient energy supply.

Environmental sustainability remains a key focus of hydropower innovation. Efforts to minimize the ecological impact of hydropower projects are driving the development of new technologies and practices. For instance, sediment management techniques are being refined to prevent the buildup of sediment in reservoirs, which can reduce storage capacity and affect water quality. Innovative fish passage solutions, such as fish elevators and bypass systems, are being implemented to facilitate the movement of aquatic species and maintain biodiversity.

Community engagement and social considerations are also integral to the success of hydropower innovations. Involving local communities in the planning and implementation of hydropower projects can lead to more sustainable and accepted outcomes. By addressing the needs and concerns of affected populations, hydropower projects can foster positive relationships and ensure that the benefits of renewable energy are shared equitably.

The future of hydropower technology is being shaped by ongoing research and development efforts. Emerging technologies, such as tidal and wave energy, are exploring new ways to harness the power of water. While still in the early stages of development, these technologies hold the potential to expand the scope of hydropower and provide additional renewable energy options. Continued investment in research and innovation is essential to unlocking the full potential of hydropower and driving the transition to a sustainable energy future.

Innovations in hydropower technology are enhancing the efficiency, sustainability, and versatility of this vital renewable energy source. By embracing new technologies and practices, the hydropower sector can continue to play a crucial role in meeting global energy needs while preserving the environment and supporting communities. Recognizing and valuing the potential of hydropower innovation is essential for fostering creativity, collaboration, and positive change in the energy landscape and beyond.

Real-World Examples of Hydropower Success

Hydropower has long been a beacon of renewable energy success, with numerous projects around the world demonstrating its potential to provide clean, reliable, and sustainable electricity. These real-world examples of hydropower success offer valuable insights into the diverse applications and benefits of this technology, as well as the challenges and solutions that have emerged along the way.

One of the most iconic examples of hydropower success is the Itaipu Dam, located on the Paraná River between Brazil and Paraguay. Completed in 1984, Itaipu is one of the largest hydroelectric power plants in the world, with an installed capacity of 14,000 megawatts. The dam provides approximately 75% of Paraguay's electricity and 15% of Brazil's, highlighting its critical role in the energy infrastructure of both countries. Itaipu's success is attributed to its strategic location, which takes advantage of the river's natural flow and elevation drop, as well as the collaboration between Brazil and Paraguay in managing and operating the facility. The project has also implemented extensive environmental and social programs, including reforestation efforts and community development initiatives, to mitigate its impact and enhance its benefits.

Another notable example is the Three Gorges Dam in China, which stands as the world's largest hydropower project in terms of installed capacity, boasting 22,500 megawatts. Situated on the Yangtze River, the dam plays a vital role in flood control, navigation, and electricity generation for China's rapidly growing economy. The Three Gorges Dam has significantly reduced the country's reliance on coal, contributing to a decrease in greenhouse gas emissions and air pollution. However, the project has also faced criticism for its environmental and social impacts, including the displacement of millions of people and changes to the river's ecosystem. In response, the Chinese government has implemented measures to address these challenges, such as investing in ecological restoration and providing support for displaced communities.

In Norway, hydropower accounts for over 90% of the country's electricity production, making it a global leader in renewable energy. The country's mountainous terrain and abundant water resources provide ideal conditions for hydropower development. One of the standout projects is the Ulla-Førre hydropower complex, which consists of several interconnected power plants and reservoirs. With a total capacity of 2,100 megawatts, Ulla-Førre is Norway's largest hydropower facility and a key contributor to the country's energy security and sustainability. The project exemplifies Norway's commitment to balancing energy production with environmental stewardship, as it incorporates measures to protect local ecosystems and maintain water quality.

The Lesotho Highlands Water Project in Southern Africa is a unique example of hydropower success that combines energy generation with water resource management. The project involves the construction of a series of dams and tunnels that transfer water from the highlands of Lesotho to South Africa, providing a reliable water supply for the region's growing population and industries. In addition to water transfer, the project generates electricity for Lesotho, contributing to the country's energy independence and economic development. The Lesotho Highlands Water Project demonstrates the potential for hydropower to address multiple challenges simultaneously, including water scarcity, energy access, and regional cooperation.

In Canada, the James Bay Project in Quebec is a testament to the scale and ambition of hydropower development. This massive undertaking involves the construction of several dams and power stations on the La Grande River, with a total installed capacity of over 16,000 megawatts. The James Bay Project supplies electricity to millions of people in Quebec and the northeastern United States, showcasing the potential for hydropower to meet large-scale energy demands. The project has also prioritized environmental and social considerations, including the establishment of agreements with Indigenous communities to address their rights and interests.

The Aswan High Dam in Egypt is a landmark hydropower project that has transformed the Nile River and the surrounding region. Completed in 1970, the dam provides electricity, irrigation, and flood control for Egypt, supporting the country's agricultural and economic development. The Aswan High Dam has enabled the expansion of arable land and improved water management, contributing to food security and poverty reduction. However, the project has also faced challenges related to sedimentation, water quality, and the displacement of communities. To address these issues, the Egyptian government has implemented measures such as sediment management and community resettlement programs.

In Bhutan, the Chhukha Hydropower Plant is a shining example of how hydropower can drive economic growth and development. Commissioned in 1986, the plant has an installed capacity of 336 megawatts and is a major source of revenue for Bhutan through electricity exports to India. The success of the Chhukha project has paved the way for further hydropower development in Bhutan, positioning the country as a regional leader in renewable energy. The project has also contributed to Bhutan's commitment to maintaining carbon neutrality and preserving its natural environment.

The Hoover Dam in the United States is an iconic symbol of hydropower success and engineering prowess. Completed in 1936, the dam spans the Colorado River between Arizona and Nevada, providing electricity, water supply, and flood control for the southwestern United States. The Hoover Dam has played a crucial role in the development of the region, supporting agriculture, industry, and urban growth. Its success is attributed to its innovative design, strategic location, and the collaboration between federal and state agencies in its construction and operation.

These real-world examples of hydropower success highlight the diverse applications and benefits of this renewable energy source. From large-scale projects that power entire regions to smaller initiatives that support local communities, hydropower has demonstrated its potential to provide clean, reliable, and sustainable electricity. However, these examples also underscore the importance of addressing environmental and social considerations, as well as the need for collaboration and innovation in the development and management of hydropower projects.

The continued success of hydropower depends on the ability to balance energy production with environmental stewardship and social responsibility. By learning from these real-world examples and embracing new technologies and practices, the hydropower sector can continue to play a vital role in meeting global energy needs while preserving the environment and supporting communities. Recognizing and valuing the potential of hydropower is essential for fostering innovation, creativity, and positive change in the energy landscape and beyond.

Chapter 5: Biomass and Bioenergy Basics

What is Biomass? An Introduction

Biomass is a versatile and renewable energy source derived from organic materials, including plant and animal matter. It has been used for centuries as a primary energy source, from burning wood for heat to utilizing agricultural residues for cooking. In recent years, biomass has gained renewed attention as a sustainable alternative to fossil fuels, offering the potential to reduce greenhouse gas emissions and promote energy independence. Understanding the fundamentals of biomass, its sources, and its applications is essential for those interested in exploring this renewable energy option.

At its core, biomass is any organic material that can be used as fuel. This includes a wide range of resources, such as wood, agricultural crops, crop residues, animal manure, and even organic waste from households and industries. The diversity of biomass sources makes it a highly adaptable energy option, capable of being harnessed in various forms and applications. The key to biomass's energy potential lies in its ability to store solar energy through the process of photosynthesis, which converts sunlight into chemical energy within plant cells.

One of the most common forms of biomass is wood, which has been used for millennia as a source of heat and energy. Wood can be burned directly for heating and cooking or processed into wood pellets and chips for use in modern biomass boilers and stoves. These systems offer a clean and efficient way to utilize wood as a renewable energy source, providing an alternative to traditional fossil fuels for residential and commercial heating.

Agricultural residues, such as straw, corn stalks, and rice husks, represent another significant source of biomass. These materials are often left over after the harvest of crops and can be used as fuel for energy production. By utilizing agricultural residues, farmers can reduce waste and generate additional income, while also contributing to a more sustainable energy system. These residues can be burned directly or processed into biofuels, such as ethanol and biodiesel, which can be used for transportation and industrial applications.

Animal manure is another valuable biomass resource, particularly in rural and agricultural areas. Manure can be converted into biogas through a process called anaerobic digestion, in which microorganisms break down organic matter in the absence of oxygen. The resulting biogas, primarily composed of methane and carbon dioxide, can be used for heating, electricity generation, or as a vehicle fuel. Anaerobic digestion also produces a nutrient-rich byproduct called digestate, which can be used as a natural fertilizer, closing the loop in sustainable agriculture.

Organic waste from households, industries, and municipalities is an often-overlooked source of biomass. Food scraps, yard waste, and other organic materials can be collected and processed into energy through various methods, such as composting, anaerobic digestion, or incineration. By diverting organic waste from landfills and converting it into energy, communities can reduce greenhouse gas emissions, decrease waste management costs, and create a more sustainable waste-to-energy system.

The conversion of biomass into energy can take several forms, depending on the type of biomass and the desired end-use. Direct combustion is the simplest and most common method, involving the burning of biomass to produce heat and power. This process can be used in small-scale applications, such as residential wood stoves, or in large-scale power plants that generate electricity for the grid.

Gasification is another method of converting biomass into energy, involving the partial combustion of biomass in a low-oxygen environment to produce a combustible gas called syngas. Syngas can be used for electricity generation, heating, or as a feedstock for producing chemicals and fuels. Gasification offers several advantages over direct combustion, including higher efficiency and the ability to utilize a wider range of biomass materials.

Pyrolysis is a related process that involves the thermal decomposition of biomass in the absence of oxygen, producing a mixture of solid, liquid, and gaseous products. The solid product, known as biochar, can be used as a soil amendment to improve soil fertility and sequester carbon. The liquid product, called bio-oil, can be refined into transportation fuels or used as a chemical feedstock. The gaseous product, similar to syngas, can be used for energy production.

Biochemical conversion processes, such as fermentation and anaerobic digestion, are also used to convert biomass into energy. Fermentation involves the breakdown of sugars in biomass by microorganisms to produce ethanol, a renewable fuel that can be blended with gasoline for use in vehicles. Anaerobic digestion, as mentioned earlier, produces biogas from organic matter, offering a versatile and sustainable energy source.

The environmental benefits of biomass are significant, particularly in terms of reducing greenhouse gas emissions and promoting sustainable land use. Biomass is considered carbon-neutral because the carbon dioxide released during its combustion is offset by the carbon dioxide absorbed by plants during their growth. This closed carbon cycle helps mitigate climate change and reduce reliance on fossil fuels.

However, the sustainability of biomass depends on responsible sourcing and management practices. Unsustainable harvesting of biomass resources, such as deforestation or over-extraction of agricultural residues, can lead to environmental degradation and loss of biodiversity. To ensure the long-term viability of biomass as a renewable energy source, it is essential to implement sustainable land management practices, protect natural ecosystems, and promote the use of waste and residues rather than primary resources.

Biomass also offers economic and social benefits, particularly in rural and agricultural communities. By providing a local and renewable energy source, biomass can enhance energy security, create jobs, and stimulate economic development. The use of biomass for energy can also support sustainable agriculture and waste management practices, contributing to a more resilient and circular economy.

The future of biomass as a renewable energy source is promising, with ongoing research and development efforts focused on improving conversion technologies, enhancing sustainability, and expanding applications. Innovations in biomass processing, such as advanced biofuels and bioproducts, hold the potential to further diversify the energy landscape and reduce dependence on fossil fuels.

Biomass is a versatile and renewable energy source with the potential to contribute significantly to a sustainable energy future. By understanding the diverse sources and applications of biomass, individuals and communities can make informed decisions about its role in their energy systems. The insights gained from exploring biomass remind us of the importance of innovation, collaboration, and sustainability in the pursuit of a cleaner and more resilient energy future.

Converting Organic Waste into Energy

The transformation of organic waste into energy is an innovative and sustainable approach to addressing two pressing global challenges: waste management and energy production. By converting organic waste into valuable energy resources, we can reduce the environmental impact of waste disposal while simultaneously generating renewable energy. This process not only contributes to a cleaner environment but also supports the transition to a circular economy, where waste is seen as a resource rather than a burden.

Organic waste encompasses a wide range of materials, including food scraps, yard waste, agricultural residues, and even certain types of industrial waste. These materials are rich in organic matter, which can be broken down and converted into energy through various processes. The key to unlocking the energy potential of organic waste lies in understanding the different conversion technologies available and selecting the most appropriate method for each type of waste.

Anaerobic digestion is one of the most widely used methods for converting organic waste into energy. This biological process involves the breakdown of organic matter by microorganisms in the absence of oxygen, resulting in the production of biogas. Biogas is a versatile energy source composed primarily of methane and carbon dioxide, which can be used for heating, electricity generation, or as a vehicle fuel. Anaerobic digestion is particularly well-suited for wet organic waste, such as food scraps and animal manure, and offers the added benefit of producing a nutrient-rich byproduct called digestate, which can be used as a natural fertilizer.

Composting is another method of converting organic waste into energy, albeit indirectly. While composting itself does not produce energy, it transforms organic waste into a valuable soil amendment that can enhance soil fertility and support plant growth. By improving soil health, composting can contribute to increased biomass production, which can then be used as a renewable energy source. Composting is an effective way to manage yard waste, food scraps, and other biodegradable materials, reducing the volume of waste sent to landfills and returning valuable nutrients to the soil.

Incineration, or waste-to-energy, is a thermal conversion process that involves the combustion of organic waste to produce heat and electricity. This method is particularly suitable for dry and high-calorific waste, such as certain types of industrial and municipal waste. Modern waste-to-energy facilities are equipped with advanced pollution control technologies to minimize emissions and environmental impact. While incineration can significantly reduce the volume of waste, it is essential to ensure that the process is conducted in an environmentally responsible manner, with a focus on maximizing energy recovery and minimizing emissions.

Pyrolysis and gasification are related thermal conversion processes that offer alternative ways to convert organic waste into energy. Pyrolysis involves the thermal decomposition of organic matter in the absence of oxygen, producing a mixture of solid, liquid, and gaseous products. The solid product, known as biochar, can be used as a soil amendment, while the liquid and gaseous products can be refined into fuels and chemicals. Gasification, on the other hand, involves the partial combustion of organic waste in a low-oxygen environment to produce syngas, a combustible gas that can be used for electricity generation or as a feedstock for producing chemicals and fuels. Both pyrolysis and gasification offer the potential for higher energy efficiency and the ability to process a wider range of waste materials compared to traditional incineration.

The successful conversion of organic waste into energy requires careful consideration of several factors, including the type and composition of the waste, the available conversion technologies, and the desired end-use of the energy. It is essential to conduct a thorough assessment of the waste stream to determine the most suitable conversion method and to ensure that the process is economically viable and environmentally sustainable.

In addition to technical considerations, the conversion of organic waste into energy also involves social and economic dimensions. Engaging with local communities and stakeholders is crucial to the success of waste-to-energy projects, as it fosters collaboration and ensures that the benefits are shared equitably. Public awareness and education campaigns can help promote the adoption of waste-to-energy technologies and encourage responsible waste management practices.

The integration of waste-to-energy systems into existing waste management infrastructure is another important aspect of successful implementation. By incorporating waste-to-energy technologies into municipal waste management plans, cities and communities can enhance their waste management capabilities, reduce landfill dependency, and generate renewable energy. This integrated approach not only addresses waste disposal challenges but also contributes to energy security and sustainability.

The environmental benefits of converting organic waste into energy are significant, particularly in terms of reducing greenhouse gas emissions and conserving natural resources. By diverting organic waste from landfills, waste-to-energy technologies help reduce methane emissions, a potent greenhouse gas that contributes to climate change. Additionally, the use of organic waste as a renewable energy source reduces reliance on fossil fuels, conserving finite resources and reducing carbon emissions.

Economic benefits also play a crucial role in the adoption of waste-to-energy technologies. By creating value from waste, these technologies can generate revenue, create jobs, and stimulate economic development. The production of renewable energy from organic waste can enhance energy security and reduce energy costs, particularly in regions with limited access to traditional energy sources.

The future of waste-to-energy technologies is promising, with ongoing research and development efforts focused on improving efficiency, expanding applications, and enhancing sustainability. Innovations in waste conversion processes, such as advanced biofuels and bioproducts, hold the potential to further diversify the energy landscape and reduce dependence on fossil fuels.

Converting organic waste into energy is a sustainable and innovative approach to addressing global waste management and energy challenges. By understanding the diverse conversion technologies and their applications, individuals and communities can make informed decisions about the role of waste-to-energy in their energy systems. The insights gained from exploring waste-to-energy technologies remind us of the importance of innovation, collaboration, and sustainability in the pursuit of a cleaner and more resilient energy future.

Benefits of Biomass as a Renewable Resource

Biomass stands as a formidable contender in the realm of renewable energy, offering a multitude of benefits that extend beyond mere energy production. As the world grapples with the dual challenges of climate change and energy security, biomass emerges as a versatile and sustainable solution. Its advantages are manifold, encompassing environmental, economic, and social dimensions, each contributing to a more sustainable and resilient energy future.

One of the most compelling benefits of biomass is its potential to reduce greenhouse gas emissions. Unlike fossil fuels, which release carbon dioxide that has been sequestered for millions of years, biomass operates within a closed carbon cycle. The carbon dioxide emitted during the combustion of biomass is offset by the carbon dioxide absorbed by plants during their growth. This balance makes biomass a carbon-neutral energy source, significantly mitigating its impact on climate change. By replacing fossil fuels with biomass, we can reduce our carbon footprint and contribute to global efforts to curb greenhouse gas emissions.

Biomass also plays a crucial role in waste management, transforming organic waste into valuable energy resources. Agricultural residues, food scraps, and other organic materials that would otherwise end up in landfills can be converted into energy through processes such as anaerobic digestion and combustion. This not only reduces the volume of waste and the associated methane emissions from landfills but also provides a renewable energy source that can be harnessed for heating, electricity, and transportation. By viewing waste as a resource, biomass contributes to a circular economy, where materials are reused and recycled, minimizing environmental impact.

The economic benefits of biomass are equally significant, particularly in rural and agricultural communities. Biomass energy production can create jobs and stimulate economic development by providing new markets for agricultural products and residues. Farmers can generate additional income by selling crop residues or dedicating a portion of their land to energy crops. The development of biomass facilities, such as biogas plants and biomass power stations, also creates employment opportunities in construction, operation, and maintenance. By fostering local economic growth, biomass enhances energy security and reduces dependence on imported fossil fuels.

Biomass offers a diverse range of applications, making it a versatile energy source that can be tailored to meet specific needs. It can be used for heating, electricity generation, and transportation fuels, providing a comprehensive energy solution that can be integrated into existing infrastructure. Biomass can be converted into solid, liquid, or gaseous fuels, each with its own set of advantages and applications. For example, wood pellets and chips can be used in biomass boilers for residential and commercial heating, while biofuels such as ethanol and biodiesel can be used as transportation fuels, reducing reliance on petroleum-based products.

The use of biomass for energy production also supports sustainable land management practices. By promoting the use of agricultural residues and dedicated energy crops, biomass encourages the diversification of agricultural systems and the adoption of sustainable farming practices. Energy crops, such as switchgrass and miscanthus, can be grown on marginal lands that are not suitable for food production, providing an additional revenue stream for farmers while preserving valuable arable land for food crops. Additionally, the use of digestate, a byproduct of anaerobic digestion, as a natural fertilizer can enhance soil fertility and reduce the need for synthetic fertilizers, further promoting sustainable agriculture.

Biomass energy systems can also enhance energy resilience and reliability, particularly in remote and off-grid communities. By providing a local and renewable energy source, biomass can reduce dependence on centralized energy systems and improve energy access in underserved areas. Small-scale biomass systems, such as micro and pico hydropower or biogas digesters, can provide reliable and affordable energy solutions for rural communities, supporting local development and improving quality of life.

The environmental benefits of biomass extend beyond carbon neutrality and waste reduction. Biomass energy production can contribute to biodiversity conservation and ecosystem restoration by promoting sustainable land use practices and reducing pressure on natural resources. For example, the cultivation of energy crops on degraded lands can help restore soil health and prevent erosion, while the use of forest residues for energy can reduce the risk of wildfires and support forest management efforts. By integrating biomass into broader environmental and conservation strategies, we can enhance the resilience and sustainability of our ecosystems.

Social benefits are also an important aspect of biomass energy production. By creating jobs and supporting local economies, biomass can contribute to poverty reduction and social development. The use of biomass for energy can also improve public health by reducing air pollution and associated respiratory illnesses, particularly in areas where traditional biomass fuels, such as wood and charcoal, are used for cooking and heating. By transitioning to modern biomass energy systems, communities can enjoy cleaner air and improved health outcomes.

The future of biomass as a renewable resource is bright, with ongoing research and development efforts focused on improving efficiency, expanding applications, and enhancing sustainability. Innovations in biomass conversion technologies, such as advanced biofuels and bioproducts, hold the potential to further diversify the energy landscape and reduce dependence on fossil fuels. By investing in research and innovation, we can unlock the full potential of biomass and drive the transition to a sustainable energy future.

Biomass offers a multitude of benefits as a renewable resource, encompassing environmental, economic, and social dimensions. By understanding and harnessing these advantages, individuals and communities can make informed decisions about the role of biomass in their energy systems. The insights gained from exploring the benefits of biomass remind us of the importance of innovation, collaboration, and sustainability in the pursuit of a cleaner and more resilient energy future.

www.ingramcontent.com/pod-product-compliance
Lightning Source LLC
Chambersburg PA
CBHW072007170726

47999CB00013B/844